Fire in the Belly

Building a World-leading
High-tech Company from
Scratch in Tumultuous Times

Jerry D. Neal
with Jerry Bledsoe

 Down Home Press, Asheboro, North Carolina

ISBN 1-878086-98-7
Library of Congress Control Number: 2004114562
Printed in the United States of America

Book Design by Beth Hennington
Cover Design by Tim Rickard
Cover Art by RF Micro Devices Communications Department

Down Home Press
PO Box 4126, Asheboro, North Carolina 27204

Distributed by:
John F. Blair, Publisher
1406 Plaza Dr., Winston-Salem, North Carolina 27103

Dedication

For my wife,
Linda Stewart Neal, whose encouragement, love, and constant
support enabled me to pursue the RF Micro Devices dream;

My business partners,
Bill Pratt and Powell Seymour, who remain not only partners,
but among my dearest friends;

And all past, present, and future employees
of RF Micro Devices.

Acknowledgments

This book was made possible through major contributions from several people and also the support of our company, RF Micro Devices. Special thanks go to:

Dr. Albert E. Paladino, who loaned me his files and inspired the title, *Fire in the Belly*. Additionally, Al provided invaluable insights through multiple reviews of the book. **Linda Duckworth**, who interviewed many of the people, researched much of the material for this book, and acted as proofreader throughout the process. **Jeff Howland**, of Womble, Carlyle, Sandridge & Rice, who provided expert advice and encouragement. **Bill Pratt** and **Powell Seymour**, who tirelessly reviewed my drafts and helped verify dates, facts, and details. Through these efforts we were able to relive the past. **Bob Bruggeworth** and **Dean Priddy**, who graciously read the draft and provided valuable feedback and support. **The RF Micro Devices Communications Department**, which provided research materials and photographs and created the art for the book cover. I am proud to have them as part of my marketing team. **Suzanne Rudy**, who provided detailed financial information and dates and is a valuable partner in my acquisition work. **Kathy Adams**, my assistant, without her help, not only on this book but through my regular schedule, I would be lost. **Jerry Bledsoe**, my co-author, throughout our work, Jerry made the development of this book fun and through the process I acquired a new friend. His expertise made this book.

Contents

1. The Gift of Layoffs 9
2. Keep a Close Eye on the Deep End 19
3. In the Footsteps of Heroes 39
4. The Mysterious Disappearing Chips 53
5. Where's King Kong When You Really Need Him? 66
6. Syncopal Episodes 80
7. Go on and Shoot Up in Here Among Us 92
8. A Gun to Our Heads 109
9. The End of the Beginning 131
10. Wall Street's Darling 145
11. Spiraling Insanity 167
12. Welcome to Moscow 189
13. Revolution, Phase II 205

Afterword 215

1
The Gift of Layoffs

When dreadful, life-altering events occur, somebody with good intentions is apt to offer doubtful solace by saying, "Maybe it's all for the best. Perhaps this has happened because better things are supposed to come from it."

That advice, sound or not, delivered or not, without doubt would have proved true in the case of Bill Pratt's tumor.

When I began casting about for the beginning of the story I am about to tell, a story of one fledgling company's struggle to bring revolutionary change to a huge new industry, it all kept coming back to the growth that was diagnosed inside Bill's skull in the spring of 1988.

The tumor was benign, but life-threatening unless removed, and in April, Bill underwent surgery at Duke University Hospital in Durham, North Carolina.

I had known Bill for seven years then. He was general manager of the company where I was a regional marketing manager.

That company was the Computer Labs Division of Analog Devices Incorporated, a multi-billion-dollar, world-wide, high-tech corporation with thousands of employees and headquarters in Norwood, Massachusetts. Computer Labs had been started by two former Bell Labs engineers in Greensboro, North Carolina, to make high-speed data converters. Bill had left RCA to join the company when it had only a couple of dozen employees. The company had grown rapidly after being bought by Analog Devices in 1978, and it now occupied a modern building near Piedmont Triad International Airport on Greensboro's western edge.

I liked and admired Bill. We traveled together often with work. I had gotten to know him well and enjoyed his company. After his operation, I visited him in the hospital. He had come through the surgery satisfactorily, but it was obvious that his recovery was not going to be a quick one.

In the charged atmosphere that was prevalent at Analog Devices at the time, I only could wonder what Bill's situation forebode.

It would not be long in revealing itself.

A new person had been brought into the Computer Labs Division to plan strategic marketing. He had a patrician air about him. I will call him the Count. People often speculated about exactly what he did, but he busied himself putting together book-size reports that nobody ever acknowledged reading. The Count seemed to have few friends at the office, perhaps because of the obtrusive cologne that he favored in such quantities that his presence was always announced long before his appearance, thus allowing people to scatter before he could corner them.

The Count clearly had high ambitions, and for months I'd been seeing indications that a coup might be afoot to oust Bill. Tension between Bill and corporate headquarters had been building for a long time. He didn't play corporate politics and was too independent to suit some in the higher bureaucracy. It was difficult to challenge him, though, because since he had become general manager in 1980, the division's revenues had grown from about $3 million annually to more than $20 million.

Bill's trouble began several years earlier when he pushed to build a semiconductor foundry in Greensboro over the objections of some at corporate headquarters. He wanted the foundry to produce a new line of products for the division.

The plant eventually was approved, but the corporation's facilities unit took so long to build it that despite Bill's constant complaints and warnings the new products were outdated before they could be manufactured. The inevitable result was a lot of blame shifting and recriminations directed toward him.

When the Count assumed some of Bill's responsibilities while he was recovering from his surgery, I began to suspect that Bill's absence would provide the opportunity for his antagonists at corporate headquarters to push him out. My suspicions were confirmed when word spread through the company that the Count was going to be the new general manager, although no official announcement had been made.

I had few dealings with the Count, for which my sensitive nose was immensely grateful, but we did share an interest—my longtime hobby, ham radio. He wanted to get his ham radio license, and we'd had a few conversations about that.

A couple of months after Bill's surgery, the Count called me into Bill's office to ask some ham radio questions, and during the conversation, he told me that he was the new general manager. The way he spoke about it caused me to assume that this was something Bill had requested and approved.

I didn't relish the prospect that this presented: the Count as my new boss. I trusted Bill and enjoyed working with him, but I had serious doubts about this guy—and I certainly wasn't looking forward to traveling with him. Nonetheless, I instructed my nose to consider undertaking the difficult and seemingly impossible challenge of adjusting to that cologne anyway.

I had called Bill a couple of times while he was recovering at home to see how he was getting along, and not long after my conversation with the Count, I called again. He said that he was doing much better and planned to come back to work soon, at least for a few hours at a time to test his strength.

"What are you going to be doing?" I asked.

"What do you mean?" he said.

I was a little taken aback.

The Count had informed me that he was the new general manager, I told Bill.

"I was just wondering what your role is going to be."

"Nobody told *me*," Bill said to my stunned silence.

"Well..." I responded, fumbling to recover, "...uh, you might want to look into this."

A few days later, Bill showed up at the office to find the Count sitting at his desk, and his secretary, Lynn Garner, embarrassed and flustered. The situation, Bill later remembered Lynn telling him, was "just temporary."

Bill had a strong disposition and was not about to give up his office without a fight. For a few days that office was the scene of a real tug-of-war.

Bill would arrive in the morning and triumphantly claim it for a few hours, even though it was a less than satisfactory victory because he later said that he had to spend much of his time scrubbing the headrest of his chair trying to remove the grease the Count had deposited there, whether from his hair pomade or the scented oils from his excess of cologne Bill couldn't be sure. As soon as Bill left, his rival would seize back the office and re-grease the chair's upholstery.

The betting was not on Bill to win this battle, because it was generally conceded that unless he started wearing a gas mask to work, he eventually would be overcome by residual cologne fumes.

Bill finally offered a solution to this standoff, and it was quickly approved by top executives at corporate headquarters. The Count would be general manager, but the company would create a design center in Greensboro to come up with new products. Bill would run that, and he would answer to corporate headquarters, not to the Count.

I got the feeling that Bill, an electrical engineering graduate of

Villanova University, really was looking forward to the change. Not only would it free him of the hassles of administration, which he never really savored, and the constant head-butting with corporate superiors, it also would allow him to return full-time to his first love, engineering.

At the beginning of August, the company rented a unit at West Friendly Business Park, building 7341, module D, on Friendly Avenue near the airport. It had room for a reception area, two offices, a small conference room, and a lab. Desks, chairs, computers, electronic testing equipment, a single-page copier, and a fax machine were trundled in and set up.

Bill was given to believe that he could build a team of engineers and technicians, and the first person he wanted was Powell Seymour. He and Powell were close friends. Both had joined Computer Labs in its early days. Powell had helped his brother-in-law, Milton Misenheimer, build a vacation house next door to Bill's at Lake Norman north of Charlotte, and Bill and Powell had spent a lot of happy summer weekends there boating, fishing, and skiing. (Later Bill's daughter, Carmina, would marry Powell's son, Chris, and Bill and Powell would share a grandchild.)

Bill went to work at the new design center four months after his surgery, and Powell, who had been in charge of a team making prototype parts, followed soon afterward. The plan was for Bill to design the new devices and for Powell to get them made and tested.

At first they worked on digital converters, but Bill had gotten interested in wireless communications, which to this point largely had meant two-way radio. But cordless and cellular phones were on the horizon, although they were in their infancy then, primitive, clunky, heavy things that few people used, or even knew about. Bill was convinced that they offered huge potential, but Analog Devices had shown no interest. He thought that it should.

He began traveling all over the country and to Europe, where cell phone technology was more advanced, visiting companies that made these devices, finding out their problems and what they needed to solve them. Integrated circuits, the semiconductor chips that powered computers, had not yet been adapted for transmitting and receiving radio signals. Bill discovered that not a single company was making such chips commercially, although they would be vital to bringing cordless and cell phones to their full potential. Some even believed them to be infeasible. Bill returned to Greensboro and went to work to prove that notion wrong.

I kept up with Bill and Powell after they moved to the business park. I'd stop by occasionally to see how they were doing, what they were working on, and what difficulties they were having. Powell's wife, Bonnie, was a technician at Analog Devices. He frequently came to have lunch with her in the company cafeteria, and I'd chat with him often.

I was especially intrigued when I learned that Bill and Powell had begun working on chips for devices that operate on radio frequencies—RF, as we refer to it in the trade. Telephone and radio had fascinated me for almost as long as I could remember.

In grade school, I checked a book out of the library on Alexander Graham Bell and was so inspired by it that I later built my own functioning telephone using coffee can lids.

After fighting in the Pacific during World War II, my dad, Albert Neal, signed up for a correspondence radio repair course. Although he never got into that work, he kept all the course materials, and I devoured them as a kid. I was always taking apart cast-aside radios and putting them into working order.

At 10 or 11, I recreated Guglielmo Marconi's original experiments with the wireless telegraph, and a couple of years later, using parts foraged from my grandpa's tool shed and junk piles, I built a primitive microphone and transmitter and went on the air with my own AM radio station in Randolph County, North Carolina, where I grew up.

I would load my high-stack 45-rpm record player with Conway Twitty, Brenda Lee, and the Everly Brothers, put my microphone beside the speaker, and my dad would drive me around the country roads so I could see how far my signal was reaching.

As a teenager, I had a friend who allowed me to hang around the radio station where he worked, WHPE, and even after high school, as I was going off to study electrical engineering, I still was thinking of becoming a radio station engineer—at least until I found out how much those guys were paid.

Although I'd never had a chance to work in radio, I remained connected through my ham activities. I envied Bill and Powell for their opportunity to get into wireless technology, but I knew there was little prospect for me to join them.

The technology business had fallen into a recession following the stock market crash of 1987, but this one would be longer and deeper than others. It proved frustrating for Bill. When he tried to bring more people into the design center, he was met with hiring freezes. When he attempted to stir enthusiasm at corporate headquarters for the products he was creating, he faced indifference. By the fall of 1990, two years after Bill and Powell had moved to the design center, they couldn't even get their calls to corporate officials returned.

Analog Devices was losing money for the first time since its earliest days; cutbacks had begun, rumors of layoffs were rife. Bill and Powell were realistic about their situation. They already had begun discussing what they would do if the company dumped them, and they had little doubt that would be their fate.

The ax came in January 1991, with the arrival of a corporate vice president named Jerald Fishman, who later would become the company's president and CEO. For years, Fishman had been Bill's primary antagonist. Years later, Bill still would recall the exact words that Fishman said to him that day: "Well, Pratt, we decided we're not going to do the RF stuff, so I guess we don't need you. But you've got one last job, and that's to get rid of him (meaning Powell)."

I found out what happened later that day when the Count came into my office, trailing fumes. "Bill got just what he deserved today," he told me with obvious glee. As he left, I wanted to hold my nose—and not just because of his cologne.

Despite the insensitivity of their dismissals, Bill and Powell actually received better treatment from Analog Devices than might have been expected. The company gave each generous severance packages—the equal of a year's pay and benefits for Bill, less for Powell. It also promised Bill the rights to the handful of products he already had developed, agreed to continue paying the rent on the office until the lease expired in the summer, offered to sell the equipment to Bill and Powell at the bargain price of $70,000, and gave them a loan for that amount until they could repay it, a deal they quickly accepted.

In essence the company was setting them up in their own business, and Br'er Rabbit couldn't have been happier to have been thrown into such a brier patch.

A month later, Bill and Powell incorporated themselves as RF Micro Devices. Years later, neither could remember exactly how the name came about. They did recall that they came up with another name first—one so undistinguished that neither could remember it —only to discover that it already had been claimed by another company. The name they eventually settled on was a simple description of the products they hoped to manufacture and market.

Bill and Powell had been planning this move from the time they realized just how tenuous their situation was at Analog Devices. But they were faced with a major obstacle: neither had the money that would be required to get the company up and running. In recognition of this, they had been compiling a list of friends and acquaintances who might consider investing in a business, including some who over the years had told Bill that if he ever decided to start his own business they wanted to back him.

Before they could take on investors, however, they needed a business plan. Bill went to Office Depot, bought a $29 "Biz Plan Builder," and following its dictates, began preparing one. It was during this process that Bill called me. From the beginning, he was keenly aware that the plan had a serious deficiency. Although he and Powell could take care of design, testing, and manufacturing (they planned to farm this out to semiconductor foundries), they had nobody to handle marketing.

"I want to offer you a job," Bill told me jocularly when he called that day. "Would you be interested?"

He, of course, already knew the answer to that question. We had discussed the possibility, and he knew how much I wanted to be part of this new venture—and how well my background had prepared me for it.

Although in the beginning of my career, I had set out to work hands-on in electrical science, I had been drawn into marketing for technology companies early on, and I had done that for more than a dozen years, first with Hewlett-Packard, then with the Burroughs Corporation, before joining Analog Devices in 1981. Between Hewlett-Packard and Burroughs, brimming with optimism, I had started my own technology company to manufacture an electronic soil-moisture sensor I had designed. But that had turned into one of the most miserable experiences of my life, causing me to swear that never again would I attempt such an enterprise. Still, Bill sensed on first meeting me that an entrepreneurial spirit simmered yet in my heart.

After interviewing me for a sales job at Analog Devices, I learned years later, Bill, while approving my hiring, told the personnel director, "That guy won't stay with us. He's had his own business. He'll want to do it again."

Bill was far more prescient than I.

"How much are you paying?" I joked to Bill when he called that day.

"That's the problem," he responded with a laugh. "We don't have any money. You won't be getting any pay."

I knew that, but it was a big complication nonetheless.

Although I had done well over the years and even had accumulated a few modest investments and a little money in retirement accounts, I still couldn't meet my expenses without a regular paycheck.

But I had no doubt that Bill and Powell were onto something big, and I wanted to be part of it, wanted to get in at the beginning and help to build a company from scratch—although I had no concept at the time just how much scratching would be required.

I told Bill that I had an idea that might allow me to join them, but it would take a little while to bring it about.

Analog Devices already had gone through one wave of layoffs

because of the recession, and speculation about more job losses was rampant. I thought that my position was safe because of my sales record and relationships with major customers, but I figured that if I could get myself included in the next wave of layoffs and get a severance package similar to those Bill and Powell received, I could make do until RF Micro could acquire enough cash to start paying the three of us.

Thus I set about trying to get myself laid off.

My dad, whose advice I greatly respected, always said that if you really wanted to get results, go to the person who could make them happen, so I went straight to the top.

I didn't know Ray Stata, the founder and president of Analog Devices, but I had been in his presence at a couple of company functions, heard him speak, and been impressed. He seemed to be a decent and straightforward fellow. I called his assistant, identified myself, said that I needed to talk with Mr. Stata, and asked if I could get an appointment. Remarkably, I soon got a call saying he would see me.

I took a day's vacation, bought a ticket with my own money, and flew to Boston.

I long since had gotten over being nervous about business meetings, no matter their gravity, but I have to admit to more than a twinge of anxiety about this encounter. My future, after all, was on the line.

Stata was a man of serious demeanor, not the kind of person to joke around with, but he was extremely cordial.

"What can I do for you?" he asked.

I told him that I had been with the company for ten years, had a good performance record, had been treated well, enjoyed working there, but I had a request that was a little unusual: I wanted to be laid off.

If he was surprised or amused by that, he didn't show it.

I knew the company was going through hard times, I went on, but it seemed to me that if layoffs were necessary, it would benefit the company more to lay off a senior person such as myself, who earned more and wanted to go, and thus preserve a job for somebody who didn't want to leave.

My purpose, I told him, was to pursue an opportunity that had come my way that I might not be able to accomplish without the severance package that the company was giving to others.

"Have you thought this over carefully?" he asked.

"Yes, I have," I replied.

"Well, what you're asking would set a terrible precedent," he said, "and I can't do it."

I'm sure my face fell at those words.

"But I'll tell you what I will do," he went on, "if you'll promise you won't go back there and tell everybody."

I would never do that, I assured him.

"If you can get the people in Greensboro to lay you off," he said, "I won't reject it if it comes across my desk. But you'll have to convince them to do it."

Back at work the next morning, March 14, I sent a memo to the Count telling him that "personal concerns" had caused me to realize that this was time for a change in my career and that I was volunteering to become part of his "headcount/expense reduction" campaign, as it was being called, so long as I could get the termination package. I followed this up with a formal request to the human resources department to be laid off.

I already had told my immediate supervisor, Andy Garner, the divisional sales manager and a nice guy, about my plans.

"No, come on," he said, "you're kidding me. You don't want to be laid off."

"Yes, I do," I said, and mentioned that I really would appreciate any help he could give me.

He tried to talk me out of it at first, but soon saw my determination.

"Well, let me think about it," he said.

A week or so passed before a list announcing another batch of layoffs came down, but my name wasn't on it. I was as glum as those who got the pink slips when I called Bill to tell him about it.

A few more weeks went by, and I heard that a new wave of layoffs was coming. I went to Andy.

"Am I on that list?" I asked.

He gave me a weak smile. "This time, you are," he said. "But it's not too late to change your mind."

"I'm not changing my mind," I happily assured him.

I was joyful. I couldn't wait to tell Bill and Powell. But it still would be a few more weeks before I could join them. In the meantime, I took a day of vacation in each of the last three weeks of April and worked with them to get a head start on my job.

On my last day of work at Analog Devices—May Day—the marketing department had a farewell luncheon for me at a restaurant called Franklin's off Friendly.

Before leaving for the luncheon, I loaded my personal belongings into a cardboard box and carried it to my faithful '79 Chevy Malibu. After lunch, I shook the hands of the men, hugged the women, and told all how much I was going to miss them. Then I drove straight out Friendly Avenue to West Friendly Business Park, building 7341, module D.

By mid-afternoon, I was sitting at a desk in a little cubicle just outside

Bill's and Powell's offices, girding myself for the great battles that lay ahead.

Never before had I felt as happy and optimistic about work and my future.

The prospect of making a lot of money wasn't even on my mind. I was just thinking how much fun this was going to be.

I finally was in the radio business.

2
Keep a Close Eye on the Deep End

Naiveté surely is a common affliction for those starting a new business. So naive were we that we had no idea that what we were undertaking actually would be revolutionary, that we would end up drastically affecting the way people communicate and the way some huge corporations did business. We did, however, dream of creating a company that would become pre-eminent in its field.

But accomplishing that was a long way off at this point, and we couldn't possibly have imagined all the complications that would ensue; all the frustration, disappointment, and discouragement; all the temporary failures and utter exasperation; all the endless hours of bone-wearying toil that would be demanded just to make tiny increments of progress.

Of this I feel certain: no pessimist ever could start such a business and bring it to fruition.

A confident, positive, perhaps even an occasional rosy-eyed view—and, yes, inevitable naiveté—are necessities to pulling anybody through all the miseries—and rare bursts of jubilation—that are certain to befall those who enter such perilous depths.

When I became a partner with Bill and Powell in their exciting new venture, I thought that my biggest immediate challenge would be to find a commissioned sales force, get those people signed to contracts and fired up about the possibilities, create specification sheets and brochures about our products and services, ship out those materials—and samples—to potential customers, and begin creating relationships with companies for which we could offer solutions to their technical problems.

But I was quickly faced with stark reality.

I had come on board with two assumptions, both naive, as could have been expected.

Number one: that the money needed to launch the company would be quickly and easily forthcoming.

Number two: that the products Bill and Powell had developed, though few in number, needed only a little work to make them operable and profitable.

Money, as is so often the case, was the first problem to ensnare me. Although many people had told Bill and Powell that they wanted to invest in the company, nobody yet actually had written a check.

It was my duty, at Bill's request, to attempt to entice them into that reluctant act. I later would wonder if even Siegfried and Roy could have pulled off that trick.

I wish I'd kept a list of the creative excuses that I heard as I made those calls. I fully expected eventually to be told, "I'd love to help you guys, but the dog ate my checkbook."

With the more promising prospects, Bill, Powell, and I called on them in a group, face-to-face (all salesmen know it's always harder to say no to a real person than to a voice on the telephone, and harder still when they appear in triplicate looking so needy), but I don't think some of them would have backed away any faster if the three of us had shown up covered with leprosy lesions.

In little more than a week, we went through the entire list of potential investors without prying so much as a dime out of anybody's pocket, and it had become apparent that we were going to have to find some other means of financing the company.

From the beginning, I had been calling friends and acquaintances I'd made in business over the years to seek their guidance, and what I kept hearing was that this was a terrible time to be starting a new business, especially a company that planned to produce semiconductor chips.

Many people had lost lots of money in technology stocks in the past couple of years, and they weren't just skittish about the prospect of investing more—some could get downright ill-tempered at the very suggestion.

Naive though we must have been perceived for our enthusiasm, we not only genuinely believed that we were special and offering investors the chance of a lifetime, we *knew* we were. This was a no-brainer to us.

Analog Devices had spent a million and a half dollars developing the products that now were ours, giving us a big head start, so much so that we were virtually alone in the obviously lucrative and wide-open field of commercial radio frequency integrated circuits.

These chips were unlike anything that anybody else had to offer, and they came not only with the many years of experience and expertise that would allow the three of us to create and sell many more—Bill was a true visionary in this technology—but also with our full-bore determination to make this into a company of which people soon would be taking notice.

Why people couldn't see that by investing with us they would be getting real value with minimal risk and great promise was a mystery that I couldn't fathom.

One of the associates I called in my quest for advice had been a customer of mine at Analog Devices, Jim Collins. Seven years earlier Jim had started his own company, Applied Signal Technology Inc., in Sunnyvalle, California, and it had become quite successful. I got right to the point.

"How'd you get the funding?" I asked.

Pretty simple, he explained; he'd just gotten together a group of local investors, and they had handed it over. That was precisely the magic we'd been attempting to perform, but in a different period, and in an area far removed from northern California's high-tech paradise, Silicon Valley, where his company was situated. People there seemed to be more savvy about the possibilities of ventures such as the one we were trying to establish in an area where tobacco, furniture, and textiles had long dominated the economy.

In fact, Jim told me, some of his backers might be interested in investing with us, and if I'd like, he'd be happy to set up a meeting with them.

Buoyed with new hope, I cashed in some of my frequent flyer miles and flew off to Silicon Valley on Monday, May 13. The meeting took place in the conference room of Jim's company on Tuesday night. About ten people showed up.

I'd prepared a presentation with lots of charts from Bill's business plan that I displayed with an overhead projector. These were not people unsophisticated about technology. They asked lots of good questions and seemed genuinely impressed with our outlook.

Some requested additional information. All wanted to think about it, of course, but I thought we had a good chance with this group. I got their business cards so I could follow up later.

I had planned this as a multi-purpose trip and had arranged several other meetings. One was with an old friend, Dennis Dzura, with whom I'd worked at Analog Devices.

Dennis had become a high-tech consultant in San Jose. I'd called him for advice early on. He had arranged for me to talk with the owners of three start-up companies. Two seemed lukewarm, and I thought we had little chance with them. But the owner of a microwave electronics company expressed great interest.

I still had to organize a sales force, and I also had arranged meetings with several sales-rep groups that had been recommended to me. The first was Cain-White & Company in Los Altos, and I came to a tentative agreement with them.

Then I flew to San Diego to talk with two other groups. One of those meetings proved fruitless, but happenstance handed me a stroke of particular good fortune on the second.

I had been scheduled to see a couple of top officers at Cain Technologies, but they, for whatever reasons, weren't available. Instead, I had breakfast with one of their reps, Kevin Kelly, whom I liked immediately.

Kevin is one of the most talented salespeople I've ever met, although I wouldn't discover that until after he started working with us. Eventually, he would bring millions of dollars in business to RF Micro, but neither of us could have foreseen that at the time. Kevin seemed to think we had possibilities, though, and indicated he would take a positive report back to his group.

I flew home feeling really good about this trip. I had lined up one sales group for certain and had a good chance on a second. I thought that at least a few of the investors I had met would come through, perhaps with large sums, and that we now were on the way to getting this enterprise off the ground.

One question that potential investors kept asking was this: If this business is such a good idea, why did Analog Devices let it go?

It was a reasonable question, and one for which I had a reasonable answer, but I knew it would be much better if Analog Devices provided that information.

Bill still had ongoing conversations with Ray Stata and others at the company, and Stata now knew what the opportunity was for which I'd asked him to lay me off. I wrote to Stata explaining my problem with potential investors and asking if he would be willing to clarify the situation. He proved himself to be the gracious person I had thought him to be.

Analog Devices had chosen not to go into the promising RF field, he responded, because it required new technology for which the company had no manufacturing capability, and to provide it would demand a broader commitment than its already over-committed resources could permit. That was all he needed to say, but to our everlasting gratitude, he added more:

"The strategy for a start-up company like RF Micro Devices can be different than for an established company like Analog," he wrote. "There are a number of companies which have been quite successful in developing markets using foundry services, and there is no reason why RF Micro

Devices cannot be successful as well. Our sense is that design expertise, knowledge, and customer service will be the important factors in developing this emerging market, and with the support of a creditable manufacturing organization, there is no reason why you cannot compete in this field."

This could only boost our chances, and I would send out many copies of this letter in days to come. But by the time it arrived in mid-June, six weeks after I had started to work, we had been forced to change our strategy.

The microwave company owner who had seemed so interested in backing us had, after several long discussions, directed his beam elsewhere. My follow-up calls to the investor group to which Jim Collins had introduced me had produced only one, Robert Briggs of Los Altos, who was willing to make a commitment—and that for just $20,000, a sum we eventually had to turn down.

To make matters worse, we had been reassessing the business plan and had concluded that we had been naive about the amount that would be required to give the company a chance to reach profitability.

The original amount we had been seeking with such scant success was $750,000. But now Bill concluded—and the facts pressed us all into agreement—that we actually would need twice that much.

After all the fruitless hours I had put into the quest so far to achieve only a single commitment for $20,000, I couldn't imagine the effort that would have to be expended to coax out another $1.48 million— and I wasn't sure we'd ever have time to accomplish it.

But none of us was willing to give in to complications, or even to entertain despair. We did our best to stay upbeat. One way or another, we were convinced, we would get the money.

The Small Business Administration was an avenue we explored for a while, and after many calls and trips to Raleigh, we discovered that we might be able to get a portion of the money there if we were willing to put up our houses, our vehicles, our retirement savings, and our children's college funds as collateral—and turn over our souls to whatever perdition the government eventually chose to assign them.

During the period when I was getting all the rejections from my follow-up calls to the potential California investors, I was talking on the phone one day to Dennis Dzura.

"I don't think this is going to work," I told him. "We're just not going to be able to get the money this way."

"I saw a list of VCs in *Electronic Design Magazine* the other day," he said.

VCs. Venture capital groups. This was a specter that had been hovering over us all along, but we had been heedfully ignoring it, pretending it wasn't there.

The very term VCs sounded sinister. The only other VCs I'd ever known about were the Viet Cong.

We had heard many horror stories about these moneyed VCs, how they would take the bulk of your company and leave you clinging to a pittance, how intent they were on controlling and micro-managing everything, even if they didn't truly understand the nature of your business or the products you were producing.

A friend had told us that hooking up with a venture capital group was sort of like keeping a pet great white shark in your swimming pool. You still could get in and swim, but you'd better keep an awfully close eye on the creature occupying the deep end. And it wouldn't hurt to be an Olympic gold medalist in the backstroke either.

The list, Dennis told me, contained the names of about forty groups that invested in electronics and included contact information and the size of the funds.

"How about faxing that to me," I said.

I still have that old thermal fax, but the print has about faded away. I well remember taking it into Bill and Powell with the sorrowful news that it all might be coming down to this.

"I think we're going to have to change our strategy again," I said. "It's going to be expensive, but I'm afraid it's going to be the way we'll have to go."

The three of us had talked about this possibility before, but now we discussed it at even greater length.

We hoped to keep as much of the company as we could, of course, and we especially wanted to set its tone and control its direction. But at this point, we realized that we had no other choice.

We knew that we could not remain a small company and make it in this market that was just beginning to take off. We had to get big and do it quickly before others did—and before the really big companies geared up and moved in. We needed capital to do that.

In the end, our reasoning was elemental: it would be far better to have a small portion of something big than a big hunk of nothing.

Soon after that discussion, I went to Eckerd's Drugs and bought a Mead composition book with a mottled black cover. On that cover I confidently wrote, "Investors."

On Monday morning, June 3, I began calling the companies on the list Dennis had faxed to me. The second call I placed was to Advanced Technology Ventures in Boston. The contact was Albert E. Paladino, who, as this is being written, is chairman of the board of RF Micro Devices. But that day he was out of his office and wouldn't be back until Friday. The person with whom I spoke told me to send a business plan and perhaps he would consider it.

My fourth call was to Brantley Venture Partners in Cleveland. The contact person there, Raymond J. Rund, wasn't available, and I was again told to send a business plan, which I did. Although it would take a while, and require developments we couldn't have imagined at the time, Ray, too, would become one of our primary investors and a member of our board of directors.

I went on to call almost all the companies on the list that day to varying degrees of success, mostly non-success. Some told me that they didn't invest in semiconductor companies, or that we were too small for their consideration, or too far out of their region. I rarely got to talk to any of the contact people on the list, but a lot of the people I did talk with agreed to let me send a business plan.

Overall, my composition book, in which I was writing down all the pertinent facts, would record this as a day largely of frustration. But the reality was that on this day I actually had set in motion the company's funding—and its future—although months would pass before that would become evident.

Another of the calls I made early on that first day was to El Dorado Ventures in Menlo Park, California. I was told to fax some information about our company to Brent Rider. The following day, Rider called to turn me down. Here is what I wrote in my composition book about that call:

"6-4. Called by Brent Rider. Nice guy—good information. They like to invest close to their location. Gave me two other names—Kitty Hawk and Hickory. Drop him a note if we get money from these sources."

Kitty Hawk Capital was right here in North Carolina, in Charlotte, just seventy-five miles away, and it wasn't on the list that Dennis had sent me. I called Kitty Hawk as soon as I got off the phone with Rider and was told to send a copy of the business plan to Walter Wilkinson.

This was proving to be a particularly trying time for our sluggish, single-page copy machine, not to mention for Bill's 18-year-old daughter, Jeanne Anne, who frequently came in to help us. Making a single copy of the 46-page business plan was a painfully slow process, and we suddenly needed dozens of copies.

I'd hang up the phone and call out to Jeanne Anne, "We need two more business plans." A few minutes later, I'd hang up and call out, "Make that three." Oh, the moans and groans I'd hear, the looks I'd get.

I knew exactly how she felt because when she wasn't there I had to make the copies myself. We could have gotten them made much quicker at Kinko's, but that was costly and we couldn't afford it.

On some occasions when I needed a copy late in the day, I would go to Kinko's in spite of the cost, then rush to the Federal Express late-night counter at the airport to overnight it. In the ensuing weeks, I'd learn the schedules of every FedEx flight out of Greensboro and get to know most of the people at the late-night counter on a first-name basis.

Bill, Powell, and I were putting in long hours at this time, and would continue to do so for years to come. We'd get to the office early in the morning, take a quick break for lunch, or sometimes eat at our desks, then work until about seven when Bill and Powell would go home to have supper with their families.

Because I lived in Randolph County, too far away to make it practical to go home to eat, I usually would get a sandwich somewhere and return to work. Bill and Powell would come straight back after supper, and we'd work until nearly midnight, then go home, sleep a few hours, and return to do it all over again the next day.

Later, I remember somebody asking me how we withstood all the strain of that period, and I had to laugh. We never felt any strain. We were like kids on a ball field who hated to quit playing and come in the house at night. We were too excited. We were having incredible fun. The long hours, challenges, frustrations, setbacks, and dis-appointments were all part of the fun.

On Monday, June 10, a week after I started telephoning VCs, I got a call that really brought an adrenalin rush. It was from Albert Paladino in Boston. He wanted two more copies of the business plan for his partners, and he told me we should consider coming to Boston for a presentation at some point after his partners had a chance to look over the plans.

The following day, I got an equally exciting call from Walter Wilkinson at Kitty Hawk Capital in Charlotte. He'd read the business plan, he said, and was interested. He wondered if it would be okay for him to drive up on Thursday, two days hence, to talk about it.

I was sure we could work him in, I said, if he'd just give me a minute to check.

We were a little uneasy awaiting our first meeting with a VC, not knowing what to expect. But Walter seemed to be a pleasant fellow with perfectly normal incisors, nothing at all like one of the hulking creatures with devouring jaws from the deep end of the pool.

A graduate of North Carolina State with an MBA from Harvard, Walter had been in the venture capital business since 1973, and had founded Kitty Hawk in 1980. He had no experience with semiconductor companies and asked lots of questions. He wanted more detailed information about our plans and in-depth résumés from the three of us, all of which we promised to get to him quickly.

We were further encouraged when he called a few days later to say that he'd like for us to come to talk with his partners, and the three of us drove to Charlotte for a second meeting.

Not long afterward, Al called to give us a date for our presentation in Boston: Wednesday, June 26.

On the off-chance that our meeting in Boston might not turn out as we hoped, I wanted us to have alternatives. In all of my calling around, somebody told me about a book called *Pratt's Guide for Venture Capital*. The library at the University of North Carolina at Greensboro had a copy, and I went there and got a list of all the venture capital firms in the Boston area.

I sent faxes to every one saying that the CEO of RF Micro Devices and I planned to be in the area the following Wednesday and Thursday talking with potential investors and that we had a little free time on Thursday morning and might be able to work them in if they were interested.

One called and actually offered to see us. Another caller said he'd like to see us but his schedule was full, and I ended up talking him into having breakfast with us before his business day began. A couple of others also called to say they couldn't see us, but one did offer to review the business plan.

On Wednesday morning, Bill and I flew to Boston on tickets purchased with more of my frequent flyer miles. Although we had no way of knowing it at the time, this would be the first of what would become many sessions over many years with Albert Paladino, whom we would come to call Al.

Al and his partners were, of course, non-committal, but we thought we had a chance after our presentation. We might not have felt that way if we had known then some things that we didn't learn until later.

One was that in the previous year, Advanced Technology Ventures had reviewed more than 500 business plans, but had invested in only two companies.

We'd have been even more discouraged if we had realized just what Al and his partners actually thought about us and our presentation.

They were intrigued with our business prospects. They just didn't think much of us.

We appeared to be knowledgeable and organized, Al wrote to his partners, "BUT very laid back. A sense of 'fire in the belly' did not emerge...."

I guess he didn't spot my roll of Tums.

This became a major point with Al, however, and a potential impediment to investment. "Conservative to the point of being boring," he later wrote of us. "Good market savvy and engineering prowess notwithstanding, fire in the belly is the missing ingredient."

I was fascinated with this when I learned about it years later, because it was so profound a misperception. I couldn't imagine anybody having more passion or confidence in what we were undertaking than we did. How, I wondered, could we have come across otherwise? I only could conclude that it must have been the cultural differences between Greensboro and Boston, and the fallacies that lifestyles of varying regions tend to create.

At this point, we had not told Walter about Al, or Al about Walter. We had been advised that VCs didn't like being played against one another and could be very touchy about that sort of thing.

We wanted to keep our options open, but in a later telephone conversation with Al, he mentioned that an investment might be more appealing if we already had some interest from a local group.

Well, as a matter of fact, we did, I said, and told him about Walter and Kitty Hawk. This turned out to be a better move than we could have imagined, because Al and Walter soon began talking and agreed to proceed together.

The next step was an all-day conference in Greensboro with Al and Walter, at which the two would meet for the first time and have a private lunch.

Al really sent us scrambling on July 8, when we received from him the lengthy agenda for the meeting, which was just four days away. He and Walter expected specific explanations to matters that in some cases we hadn't even had time to think about.

We began to wonder if you ever could get enough information for these guys, no matter how much you worked. They wanted to know *everything*—and then some. And now.

Among the things they wanted to know were these:

Why had Bill chosen the products he had designed? What was their function and current status? How would they fit into the market? What would they cost to produce? And what would our pricing strategy be? What future products did we have in mind? How would we accommodate a big order if we got one?

To my dismay, marketing was an area that they found deficient in the business plan, probably because Bill had written it before I joined the company, although we had been regularly updating it. They wanted specific sales projections. They wanted to know who would be selling and how, who the competition was, who might become competition, and how we would meet their challenges.

They wanted to know the exact job positions we needed to fill presently and beyond, what the schedule was for that, how we would find these people and where, and who would do it and what they would expect to be paid.

They wanted a full explanation for the funding we had requested, and a plan for financing growth when it came. They wanted to know how big we thought the board of directors should be, who should be on it, and what we expected our role to be in governing the company.

One thing on the agenda really bothered us. They wanted us to justify why the company should be in Greensboro, where technology companies were few. Shouldn't we be in Silicon Valley, or some other area where high-tech companies were concentrated, at least in the Research Triangle of Raleigh-Durham-Chapel Hill, which was more conducive to this industry?

On this we had decided to draw a line. I was the only native of the area, but both Bill, who had come from New Jersey, and Powell, who had arrived from Florida, had lived here for more than 20 years. They loved the area, and their families were settled here.

I revered the beautiful, rocky, and forested red-dirt hills of Randolph County where I lived, and where my roots reached many generations deep. The couple of times I had moved away briefly, I had longed to return, and I had no desire ever to leave again.

The next three days were frantic as we dug for information, made projections, wrote reports, drew charts, and engaged in lengthy and deep discussions about strategy.

We were greatly relieved when we delivered Al to the airport for his 3:35 flight back to Boston on July 12, and Walter was driving home to Charlotte, but we thought we'd fared okay. Still, we had no promises, only hope.

When next we heard from Al, he wanted to send a consultant to assess our technology. This was Erik van der Kaay, the president of Antenna Specialists, a division of Allen Telecom Inc., a well-established manufacturer of wireless communications equipment, in Beachwood, Ohio.

That was fine with us and we began preparing for his visit on August 23. But before that happened, Al called to say that he, too, would be coming, as would Robert Paul, the president and CEO of Allen Telecom. We were pleased to be receiving the attentions of such distinguished visitors, and obviously knew we needed to make a good impression.

The agenda for this session was essentially a repeat of our presentation to Al and Walter nearly six weeks earlier, but with more technical information and updates on new material we had developed.

We even put on a demonstration, something I had done regularly at Analog Devices so customers could see just how certain components worked.

I hooked one of the prototype receiver chips Bill had designed to a circuit and connected it to an inexpensive scanner I bought at Radio Shack. I tuned the scanner to cell-phone bands, and we eavesdropped on conversations of unwitting participants, a couple of which turned out to be entertaining. This produced some laughs and lively conversation, but more importantly, it showed that we had a product that actually worked.

Although we didn't know it at the time, we made quite an impression on both Erik van der Kaay and Bob Paul during this visit. Erik had realized on first seeing our business plan that we had products his company could use and that we were ahead of everybody else in this field. He wanted his company to become an investor then, with a possibility of acquiring RF Micro outright down the road.

After their visit, both he and Bob Paul were even more enthusiastic. They thought our products would receive extraordinary acceptance, that we would be alone in our field for a good period, and that we were far underestimating our potential market share, something we had done not only because of our conservative nature but in an attempt not to seem unrealistic or cocky.

"If only the principals could be so fired up," Al wrote a few days later of their report.

In essence, our potential investors thought that we were doing everything exactly right—we just weren't the people who should be doing it. Already they were discussing that the company would need a less laid-back CEO and a more vigorous marketing VP.

Although Walter and Erik were ready to go ahead with the investment, Al still was holding back.

In the meantime, an unexpected development had put us in a dilemma.

One of the friends I'd called for advice during our quest for cash was Mike Pawlik. Mike once had worked for a company called Gigabit Logic, which was partially owned by Analog Devices until it was absorbed by other companies. All of us had worked with him during that period and liked him a lot. Now he was vice president of marketing for Burr-Brown Corporation in Tucson, Arizona, a company about the size of Analog Devices, maybe a little bigger, and a major competitor of our former employer.

When I told Mike about all the problems we were having raising money, he said, "Why don't you send me a business plan, let me have a look at it."

Not long before the visit by Erik van der Kaay and Bob Paul, he called to say that Burr-Brown might be interested in funding us. He also asked for more copies of the business plan, sending our faithful but lagging copy machine once again into overtime.

When Mike called back it was to tell us that Burr-Brown executives had discussed our plan and wanted to know if we were willing to fly out at their expense to talk about it. The receding remains of my frequent flyer miles cried out with relief as I assured him that we would.

On Monday, August 28, five days after our assessment by Erik and Bob Paul, Bill, Powell, and I flew to Tucson. Burr-Brown put us up at the Embassy Suites Hotel, and Mike met us for dinner that night. We had a great time telling old tales, laughing, and talking about the prospects to come.

The following day, while Powell met with the manufacturing people, Bill and I laid out our plan for Burr-Brown's top executives and fielded their questions. They seemed quite receptive.

In the afternoon, the three of us gathered with the company's technical staff, and we put on a demonstration of one of the chips Bill had designed. Mike assured us that everything had gone well, and we flew home on Wednesday with new possibilities—and uncertainties— buzzing in our heads.

We knew that Burr-Brown's executives might be salivating over the possibility of taking in Analog Devices' castaways and turning them into an asset, especially if we succeeded as we expected to, and we thought that gave us an edge. We also believed that we would get a better deal from Burr-Brown than from the VCs. But, of course, we had no way of knowing that, or what the top officers and directors might decide.

All through September we were kept hopping, talking constantly on the phone to our prospective investors, sending rafts of information and updates back and forth to both the VCs and Burr-Brown.

We felt certain that an offer would be coming from the VCs, and we expected it to be onerous. We just didn't know how onerous until Walter faxed it to us on September 20.

On agreeing to the term sheet, we would receive a bridge loan of $75,000. On signing the final contracts, we would get another $325,000. To get the remaining $1.1 million, we would have to: complete the first seven products and produce "satisfying" production runs; have an unspecified number of key customers design our products into their devices and order an unspecified number of units over a three-year period; define and design an unspecified number of new products; define the business plan to precisely determine the company's ultimate size and speed of market penetration; hire a CEO, a senior design engineer, a marketing assistant, and a financial officer; determine the location of the company when it outgrew present quarters; and purchase a $1.5 million life insurance policy on Bill made payable to the investors.

Powell and I couldn't resist pointing out to Bill that if he got run over by a dump truck after we received the first $400,000, the investors could make a quick $1.1 million profit on him alone.

The investors wanted a board of directors that gave them complete control. It would consist of no more than seven people, three of which would be investors, two of which could be founders, and two outsiders, one chosen by each group.

In case the company had to be liquidated for whatever reasons, all money would go to the investors until their investment was recovered, plus interest.

Most onerous of all was that for this $1.5 million investment, the VCs would get 60 percent of the company. Another 4 percent would be set aside for employee stock options, and we would be left with 36 percent..

Three days later we responded with a three-page letter gently and politely suggesting that a partnership actually should be a partnership. Why not equal ownership, 48 percent for founders, the same for investors, and 4 percent for employee stock options? And shouldn't the board of directors consist of equal representation, three founders, three investors, one outsider chosen by agreement?

We had quibbles, too, with the stipulations placed on receiving the bulk of the financing. Shouldn't they be goals instead of conditions?

The answer to this was quick in coming. We got it just two days later. Essentially, it was no to all the major points, with a few minor concessions.

By this time, Walter and Al had become suspicious of our travels. "Jerry left yesterday," Walter faxed to Al on the day he sent us the revised term sheet. "Don't know whether or not it's to meet other VCs as you suspected last time."

Bill was leaving that day for a West Coast trip, so it was a few days before we got together to consider a response. Before we could formulate one, on October 3, we received yet another revised term sheet. Apparently, a new set of lawyers had gotten hold of it, and had tightened it more in favor of the investors.

Ironically, on that same day, we got a letter of intent from Burr-Brown to provide our funding.

It was a pretty straightforward arrangement without all the complexities of the VCs' proposal. Burr-Brown would give us $375,000 on signing the final agreements, another $250,000 two months later, and $125,000 per quarter until the full $1.5 million investment was reached. No stipulations had to be met, other than that Burr-Brown would be the exclusive distributor of our products.

For the first $750,000, Burr-Brown would get 40 percent ownership. They would get another 40 percent for the second $750,000, but we would have the option of buying back that second 40 percent share at any time at the investment price plus a 25 percent annual return.

In other words, if we did well and made money, as we fully expected to, we could use that profit to buy back the second 40 percent. We then would own 60 percent of the company and control its destiny.

With the VCs, we always would hold a minority share and never have control. We would have to depend on their goodwill to allow us to lead the company in the direction we wanted it to take.

Was there really a choice? We didn't think so. We added a few amendments to Burr-Brown's letter of agreement, signed it, and sent it back on October 14.

It was during this period that I first got a call from Ray Rund with Brantley Ventures in Cleveland. His was the fourth group I'd called on the day I began phoning VCs back in June. He now had read our business plan and wanted in on the deal. I'm not sure what I told him at first, but I must have said that we already had arranged financing, and I also must have mentioned Al and Advanced Technology Ventures, because by mid-October, unbeknownst to us, Ray was in touch with Al.

Ray also kept calling me, expressing his continuing interest, and offering advice. He kept calling even after I told him that we had accepted an offer from Burr-Brown. We should consider some VC money, too, he said. Depending on a single corporation was dangerous. Market conditions, or the company's leadership, might change, and we suddenly could find ourselves out in the cold.

Although I'd never met Ray, I got to like him through our conversations, and I was impressed with how persistent he was, and how much he wanted to be involved, even though I saw no role for him.

Meanwhile, Walter and Al were getting anxious about why we hadn't responded to their most recent term sheet, and we kept putting them off until we got the final contract from Burr-Brown. Two weeks after we returned the signed letter of intent to Burr-Brown, Bill, Powell, and I went back to Tucson to work out the remaining details. We met with the company lawyers, reviewed the proposed contract, and suggested changes. We also met with the financial people about the schedule for the actual transfer of funds.

While we were talking, they told us that they knew we needed money and would make us a loan of $75,000 to tide us over until everything was settled. We could pay it back when the investment money arrived. They even drafted a check.

Everybody told us how happy they were to have us in the Burr-Brown family, and we couldn't have been happier ourselves as we left. We actually had money in hand after months of intense and frustrating toil to that ever elusive end.

I'll never forget the chore that awaited me on our return. I set up a conference call with Walter, Al, and Erik, swallowed hard and told them, "We really appreciate the way you've been working with us all this time, but we have what we think is a better offer."

I suspect they probably weren't accustomed to getting this type of call. For a moment, there was what I took to be an astonished silence. Then I began getting it from all sides. They were, to put it mildly, somewhat irritated, especially Al.

I responded in a conciliatory manner, told them that I understood why they felt as they did, but we had a responsibility to do what we thought was best for the company, and this was it in our view.

"We haven't signed a definitive contract," I told them. "Conditions could change. So why don't we keep in touch, and if anything does change, I'll let you know."

I've long been convinced that business functions on relationships, and that making and keeping relationships is essential to success. It's never a good idea to break off relationships, no matter how circumstances might sully them.

I went to Bill and Powell and told them that I thought we needed to do something to smooth over this situation and try to stay in the good graces of Walter, Al, and Erik.

We decided to offer Walter a seat on our board of directors. I waited an appropriate time, called Walter, asked if I could come to talk with him, and he agreed. Flattered though he was, he said, his fund's rules didn't allow him to accept positions on boards of companies in which his firm had no investment.

I told him that I hoped we could work together again. Perhaps he could serve as consultant, or in some other role, and we parted on friendly terms. That simple act of courtesy would end up saving our company.

• • •

The first thing we did with the loan we got from Burr-Brown was to start employing people. Linda Duckworth, a young woman who had studied business at Elon College, left an insurance company to became our first hire on November 12. Her job was to do whatever needed to be done, and she began working with me on the product specification sheets that we had to get to Burr-Brown for its catalog.

To this point, we had used a part-time bookkeeper to record our expenditures, but with investment money coming, we needed somebody to handle finances full-time. The first person to come to mind when we talked about this was William A. Priddy, who was called Dean, a nickname derived from his middle name, Aldeen.

Like me, Dean was a Randolph County native, but he was from the small town of Liberty on the far eastern side of our big county, while I was from the modestly mountainous western section. Dean had a master's degree in business administration from the University of North Carolina at Greensboro, and Bill had hired him straight out of college to work in the finance department at Analog Devices. Dean had since left the company to work for Sarah Lee.

Powell had kept up a friendship with Dean, who was an avid saltwater fisherman and frequently brought fish fresh from the coast to Powell and his family. Later, Dean would remember that when Powell called to offer him a job, he thought Powell was calling to find out when he was going to bring more fish. Dean was scheduled to start work on Monday, December 2.

But on Saturday, November 30, I got a call at home from Mike Pawlik in Tucson.

"Jerry," he said, without the usual cheerfulness in his voice, "we've got a problem."

"What kind of problem?"

The president and CEO of Burr-Brown, who had been in charge of the negotiations for our investment deal, was missing, Mike told me.

"Missing?" I said. "What do you mean?"

"He's just gone," Mike said. "Took off."

"Do you have any idea where he's at?"

He could be in Mexico, Mike said. They just didn't know.

"What brought this about?" I asked.

"We don't know that yet either," Mike said.

Then he got to the bad news. The board of directors had decided to put a hold on all the company's unsigned contracts and negotiations until they could get to the bottom of this.

Mike sure knew how to take the fun out of a weekend. We were only days away from signing the final contracts at this point, and for the first time since we had started down this long and trying road, I felt a genuine sense of dread.

Mike apologized, said he hoped this all would work out, and that he would keep us informed of what was going on.

The calls I made to Bill and Powell that day were glum indeed. Our worry was that if this brought about the collapse of our agreement, and if Walter, Al, and Erik had moved on to other investments and were too miffed to take up where we had left off, then we might have run out of time to start all over again.

If we didn't get more money soon, we would have to let Linda and Dean go, and Dean hadn't even started to work yet. Also the three of us were at the end of the severance money we'd gotten from Analog Devices. If we took too long to find new investors, we might be forced to admit failure, terminate the whole thing, and go in search of jobs ourselves.

Clearly, this situation was beyond our control. We could do nothing but wait to see what happened. Little was to be gained from dwelling on the negative, although it was hard to ignore.

Dean came to work on Monday morning for the first time, and we all went about our business as if nothing unusual was going on. It wasn't until a few days later that we heard again from Mike.

"Jerry," he said, "it's bad news."

Burr-Brown's board had decided to cancel all pending contracts negotiated by the missing CEO. (We never did find out if he ever was located, or what prompted his disappearance; our minds were too fixed on other matters.)

Bill, Powell, and I huddled in Bill's office to try to decide what to do. We had to let our two employees know about this situation, we quickly agreed, so they could decide whether they wanted to continue under the circumstances.

Bill said that he had a little money put away that he would let the company have to pay Dean's and Linda's salaries until we could determine whether we had any real chance to find investors.

We thought we might be able to shame Burr-Brown into granting us another loan to help us until we could work out a deal with somebody else, and I agreed to call Mike with that proposition. Mike and others in management at Burr-Brown felt really bad about this and realized what a bind it had put us in. As it turned out, they did talk the board into agreeing to let us have another $75,000 loan.

Bill called Dean into his office to tell him about our dire situation. I informed Linda and told her that if she were uncomfortable about staying and wanted to begin looking for another job, we would understand. As it happened, a major insurance company she had applied to before coming to work with us had called to offer her a job. She talked to her father, a partner in a local CPA firm, and although he didn't tell her what to do, his advice left her with a dilemma. "The sure path," he said, "would be to accept the job with the established company."

"What do *you* think I should do?" she asked when next we talked.

I wouldn't presume to advise her, I said, but I did think that we still had a chance to get the money, and that we eventually would make it as a company, and she would have the opportunity of being in at the beginning of something that could be big. In the end, she chose to stick with us, and I've always admired her for that spunk and dedication.

The reason I was able to express confidence about getting the money wasn't just due to my optimistic nature, but because I already had made the call I dreaded most after our bad news—the one to Walter Wilkinson at Kitty Hawk Capital.

"I think there's a chance we can put this deal back together, if you guys are still interested," I told him, struggling to put the best face on dire straits. Our other funding source wasn't working out exactly as we had hoped, I explained, and he gracefully didn't press for details.

He was interested, he said, but he didn't know about the others. He'd have to check with them.

Well, even if they weren't, I told him, we had somebody else who was, and I told him about Ray Rund, not realizing that he and the others had long been aware of Ray and our arrangement with Burr-Brown. I thought that Ray's interest might spur Walter, Al, and Erik to return to the fold.

Walter called back later to say that he had talked with Al and Erik, and they still were interested and would be willing to include Ray. They'd try to put the deal back together, Walter said, but we'd face a whole new round of checking and would have to supply a lot more financial information.

That trying process went on for more than a month, but at 3:05 p.m. on January 8, 1992, almost a full year since Bill and Powell had been laid off by Analog Devices, our little thermal fax machine, which sat between the faithful copier and the essential coffee maker, began beeping, and we gathered around as it spewed out the nine-page term sheet that Walter was sending.

All we had to do was sign it, return it, and we would get a $100,000 bridge loan to carry us until the final contracts were worked out. The terms were similar to those we had found so onerous four months earlier, although we would get half the money on signing and the rest later. But we were in no position to challenge.

RF Micro Devices was going to get a chance to be a real company, after all, even if we never again would control it completely.

Some might consider what we went through to get to this point an ordeal, but I didn't see it that way. I've long believed that human beings are designed for struggle, and that only through struggle can they achieve genuine satisfaction and happiness.

When I look back now at those hectic and often distressing months when just the three of us, Bill, Powell, and myself, were working so hard and dreaming so big, I still think of that period as the most exciting and fun time of my entire career.

We also were well aware that the struggle was just beginning. Now we had to make that dream of creating an important and lasting company a reality.

But at least, finally, we'd be getting paid.

3
In the Footsteps of Heroes

A few months after I went to work with Bill and Powell, I finally found time to take my dad to see the company one Saturday when nobody was working.

Dad was in his 70s then, a man of succinct observations. It didn't take long to show him around our entire 1,854 square feet.

He was fascinated with anything technical, so I demonstrated some of the lab equipment and showed him the sample chips Bill and Powell had made.

I went on at great length about what a big market cell phones were going to provide, and how we soon would be selling millions and millions of these chips.

As Dad looked around at this tiny operation, I could see in his face that he might be wondering if his only son had become a little bit delusional.

But if that were so, he was too polite to mention it.

Instead, he asked, "Who are you going to be competing with?"

"Companies like Hitachi, Motorola, AT&T," I replied, calling off a few of the giants to impress him.

"Oh..." he said, looking around at the entirety of RF Micro Devices and nodding as if his suspicions had been fully confirmed. "That's good."

At that time, I suppose a lot of people might have compared our ambitions with our resources and concluded that, if not delusional, we were at least audacious. And I guess we would have had to plead guilty.

After all, we had no certainty then of getting the company funded, or of surviving longer than a few more months. And all we had to offer, besides our abilities, our enthusiasm, and our commitment to endless work, were seven miniature products, all of which would fit into a small match box with plenty of room to spare.

These were the microchips that Bill had designed for Analog Devices.

We had great faith in them, as well as the others we planned to produce, because we knew they were the next step in the evolution of technology that was allowing what was once inconceivable to become reality—the ability to communicate instantaneously by voice, written word, and eventually even by vision, with anybody, anywhere, at any time.

We were just taking our place in a long line of dreamers who, one step at a time, had come an incredible distance toward making that possible.

This technology reached all the way back to those who first saw potential in electric charges, lightning being the most obvious. By the early 1700s, others had found ways of producing more modest charges and were learning that they could be controlled.

In 1729, Stephen Gray, an English chemist, discovered that some materials, such as copper, were good conductors of electric charges, while others were insulators. He was the first to transmit electric current—energy produced by flowing electrons—over a wire. (The word *electron*, incidentally, comes from the Greek word for *amber*, the petrified tree resin which ancient Greeks discovered would attract light objects when rubbed, the effect of static electricity.)

Just 24 years after Gray's experiments, the idea of using electricity to send messages was proposed in *Scots Magazine*, but it would remain only an elusive concept for nearly a century.

In 1800, an Italian physicist, Allesandro Volta, made a huge technological leap by inventing the battery, thus providing a continuous electric current for the first time and getting his name forever memorialized as the means for measuring it.

A Danish physicist, Christian Oersted, discovered that an electric current creates a magnetic field and demonstrated it at the University of Copenhagen in 1820.

The following year, an English scientist, Michael Faraday, reversed Oersted's experiment, wrapped a wire around a magnet, and discovered that it produced a faint current, thus creating the electromagnet, which then was called a dynamo. The dynamo was the basic element of a generator, a device that allows mechanical energy, such as falling water or wind, to be transformed into electricity, although effective generators would be many years in the making.

By 1830, following Faraday's lead, an American professor, Joseph Henry, had created the first efficient electromagnet and used it to transmit the first practical signal by electricity.

Henry also would assist Samuel Morse, a painter and professor, in making the great breakthrough into electronic communications, the telegraph, which Morse invented in 1837.

• • •

Morse's device was simple. He used a battery to create a current and a switch, or key, to turn it off and on, sending pulses of energy along a line connected to a receiver. He also devised a code of dots and dashes representing letters, numbers, and punctuation marks so that messages could be sent. A short burst of power was a dot, a longer one a dash.

By 1844, the first telegraph line connected Washington to Baltimore, and Morse himself sent the first message over it: "What hath God wrought!"

What indeed?

The Western Union Company was organized in 1851, and within a decade, it had strung telegraph lines all across the country, usually following railroads. Railroad companies and newspapers became the earliest users of this fast, new means of spreading information.

Morse's first telegraph employed a mechanical pencil, driven by an electromagnet, to mark dots and dashes on paper. By the 1850s, resonators had been added to telegraph keys, making the signals audible, and telegraph operators were trained to translate messages by sound.

The telegraph made it possible for individuals to quickly get messages to others in distant spots, and Western Union rolled in money. The country was connected by almost instantaneous news, with all of its profound consequences. The cost of sending telegraphs also produced a blessed brevity not since known in mass communications.

Even so, telegraph lines were so busy that by the 1870s a new ailment had appeared: telegrapher's paralysis. Computer operators today know that as carpal tunnel syndrome.

Telegraphy was in its heyday in 1871 when one of my childhood heroes, Alexander Graham Bell, born in Scotland, came to this country by way of Canada. A teacher of the deaf, he was interested in sound and telegraphy. While experimenting to find ways to send multiple messages over telegraph lines at the same time by changing the pitch of signals, Bell realized that he heard the twanging of a nearby clock spring over his line.

If the sound of a clock spring could pass along a line, why not a voice? All it would take was a diaphragm at each end of the line to vibrate the voice along the current and retrieve it at the other end.

Another person, Elisha Gray, was working on the same idea at the same time. Bell beat him to the patent office only by a few hours. Within three weeks, using ideas in Gray's patent application but not his own, Bell had contrived the contraption that would become known as the telephone, from the Greek words, *tele* (from afar) and *phone* (voice).

On March 10, 1876, in Boston, Bell made the first telephone call to his assistant, Thomas Watson, who was in an adjoining room. If he had given more thought to history, his message might have been as awe-filled as Morse's, or at least inspirational. But he only came up with the practical and mundane, "Mr. Watson, come here, I want you!"

I've never seen mention of a reply by Watson, so I only can assume from this message that Bell was a little uncertain about whether this contraption actually would work, and that what he wanted was to find out if Watson had heard anything. But if instead he was seeking advice on a rousing and truly memorable first telephone message, his timing was a little off.

The first outdoor telephone line, three miles long, was erected in Boston in 1877, the same year the Bell Telephone Company was formed. A year afterward, the first commercial telephone exchange opened in New Haven, Connecticut. The exchange allowed calls to be switched among numerous subscribers. That had to be done by hand, creating the job of telephone operator.

Local phone networks spread quickly in cities across the country, with reception greatly aided by Thomas Edison's invention of the carbon microphone in 1888 (as a kid, I would, if you'll pardon the term, jerry-rig such a microphone of my own). Bell's company wasn't alone in building telephone networks. Elisha Gray's Western Electric became an early competitor, but the Bell Company would buy controlling interest in it in 1882.

In 1879, the first telephone numbers were assigned, and a year later a mortician in Kansas City invented the dial telephone because he thought that an operator was forwarding calls intended for his funeral home to a competitor. But not for decades would dial telephones become the norm.

Into this picture comes another of my childhood heroes, Guglielmo Marconi.

In 1886, working from British physicist James Clerk Maxwell's theory of electromagnetic waves, published in 1873, a German physicist, Heinrich Hertz, created the first artificial radio waves in a classroom, proving Maxwell's theory. Hertz demonstrated that radio waves had the same qualities as light, could be reflected and refracted, and moved at the same speed as light. The means of measuring radio waves, or frequencies, would come to bear Hertz's name.

Using Hertz's discoveries, Marconi theorized that telegraph messages could be sent by radio waves, and by 1895, he was doing just that for short distances, using an electromagnet, spark-gap transmitter. He became the first to send a radio message to a ship at sea, 30 miles away. And two years after founding Marconi's Wireless Telegraph

Company in London in 1897 (his mother was British), he established commercial radio communications between England and France.

Radio waves could reach only for certain distances because they traveled in straight lines and the earth's curvature caused them to shoot into space. Marconi discovered that by grounding his transmitter, the signals would follow the earth's surface. He demonstrated it on December 12, 1901, when he and his assistant, George Kemp, received the first radio signal transmitted across the Atlantic from Poldhu, Cornwall, England, to Signal Hill in St. John's, Newfoundland.

Thus was born radio and all the wonders that eventually would spring from it. The marvelous devices that radio would produce would change human existence forever—and so much for the better. They also would entrance me throughout my life.

Navies and ship owners were quick to see the potential of the wireless telegraph. For the first time, ships at sea could be tracked, be warned of storms and other dangers, and call for help when needed. A universal code for distress—SOS—was created in 1906, although it wasn't adopted by the United States until 1912. (Incidentally, the first SOS signal in this country was received in my home state, at Cape Hatteras on the Outer Banks, in August 1909. It came from the SS *Arapahoe*, which was foundering near Diamond Shoals in the infamous "Graveyard of the Atlantic." The *Arapahoe* was saved and not long afterward was involved in the second SOS in the country, this time receiving the signal from the SS *Iroquois*.) The wireless telegraph played a major role in rescuing the survivors from the sinking of the *Titanic* in April 1912. It would play an even more important role in the First World War.

Marconi's success created what might be considered the first high-tech stock market rush. Numerous companies sprang up to exploit the wireless telegraph and other commercial possibilities that radio waves offered. Although most had few assets and little chance of profit, that didn't stop investors from driving stock prices to such outrageous highs that in 1907 *Success Magazine* published a series of articles called "Fools and Their Money/The Wireless Telegraph Bubble."

But even then scientific developments were taking place that would offer huge market potential for some companies.

In 1904, a British scientist, John Fleming, used Thomas Edison's only scientific discovery—that an electric current didn't require a wire to conduct it but could move through a gas, or a vacuum—to create a vacuum tube, now known as a diode, that would change an alternating current into a direct current.

Two years later, a playboy tinkerer named Lee De Forest made a major improvement on Fleming's tube, turning it into a triode. He, however, called it an audion. Some of those who first saw it demonstrated at a lecture in Brooklyn referred to it as a "queer little tube" and had no idea of its prospects.

Whereas Fleming's tube had but two elements—a filament that heated and glowed when connected to a power source, and a positively charged metal plate—De Forest added a third: a negatively charged metal grid between the filament and the plate.

When the filament was heated, it created an electrical field, releasing huge quantities of negatively charged electrons that raced to the positively charged plate. As the electrons washed over the grid, they picked up the signal being fed to it and amplified it many times over before flowing from the plate on an outgoing line.

Although few people realized its value at the time, De Forest's vacuum tube would revolutionize communications, eventually making possible radar, commercial radio and television broadcasting, and early computers. (While some might question the beneficial aspects of much of the programming that radio and TV eventually would produce, I still look fondly back on those nights of my childhood when our family gathered around the console radio with its mesmerizing green-glowing tubes to listen to *The Jack Benny Show, Our Miss Brooks, Inner Sanctum,* and other memorable shows.)

But the first major use of De Forest's vacuum tube was to make transcontinental telephone calls possible. Until then, long-distance calls had been limited because the signal eventually faded away. The Bell System bought the rights to De Forest's invention and strung a telephone line from New York to San Francisco with De Forest's tubes fitted along the way to boost the signal.

On January, 25, 1915, Alexander Graham Bell in New York made the first ceremonial transcontinental call over that line to his old friend Thomas Watson in San Francisco.

In the nearly four decades that had passed since he made the first phone call, Bell still hadn't been able to come up with anything memorable to say.

"Mr. Watson," he repeated, "come here, I want you!"

"Whaaaa?" said Watson.

No, I'm just kidding. He didn't say that. He actually replied, "It will take me five days to get there now."

Although it used too much power; created far too much heat; and was fragile, short-lived, and often unreliable, the vacuum tube with all of its refinements would reign for nearly half a century. And it would be the Bell System that would come up with its replacement.

● ● ●

At the end of World War II, Bell Laboratories, the research division of the American Telephone and Telegraph Company, which had become the parent company of the Bell System in 1899, formed a team of scientists to find a solution to the problems of the vacuum tube, which by then served in many capacities in communications devices.

The team was organized by Mervin Kelly, Bell Labs' director of research, who in the 1930s had become intrigued with a group of materials called semiconductors, which he thought might hold the future for electronics. Semiconductors are crystal elements such as silicon or germanium, which, when slightly altered, can carry and control an electric current.

Kelly chose a brilliant young scientist named William Shockley to head the team of physicists, chemists, and engineers who were to look into semiconductors as more efficient alternatives to vacuum tubes. Shockley soon designed a small device using silicon, an element common to sand and glass, but it didn't work. He assigned two physicists on his team, Walter Brattain and John Bardeen, to figure out why.

Shortly before Christmas 1947, Brattain and Bardeen revealed a small device made of plastic, germanium, and gold foil with three electrical connections. The device had the essential elements of the vacuum tube and performed the same functions using only a minute fraction of the power.

Shockley was said to be furious that Brattain and Bardeen hadn't kept him informed of their progress, and in four weeks, he designed a better device.

On June 30, 1948, Bell Labs revealed to the world the invention it called the transistor, although the world initially took little notice. Stories vary as to how the transistor got its name, but there can be no doubt of its effect. It would change electronics as surely as the vacuum tube did.

Shockley clearly saw its possibilities and left Bell Labs to start his own semiconductor company in Palo Alto, California, where he had grown up. He hired some of the brightest scientists and engineers available to work for him, but many soon left because of his overbearing personality. The deserters formed such successful nearby companies as Fairchild Semiconductor and later the Intel Corporation, marking the beginning of Silicon Valley.

Shockley received the Nobel Prize for the transistor in 1956, but his company never found success. He took a teaching position at nearby Stanford University and eventually became a highly controversial figure because of his theories on genetics and intelligence. Brattain and Bardeen, who also won Nobel Prizes with Shockley, remained at Bell Labs until also leaving to teach.

The transistor was far smaller, far more efficient, and far more reliable than the vacuum tube. Called a solid-state device because of its crystal composition, it produced little heat and could be used as an amplifier, a detector, or a switch. It also allowed bulky equipment to be greatly reduced in size.

Transistors were first used primarily in telephone and military equipment in the United States, but their commercial consumer potential didn't go unnoticed by two Japanese engineers, Masaru Ibuka and Akio Morita. They formed a company called Sony Electronics to manufacture tiny, transistorized radios and helped to make the marvels of the transistor available to everybody.

At the same time that Shockley and his crew were creating the transistor, radio and telephone finally merged commercially in the mobile phone.

Transmitting voice by radio waves was a natural outgrowth of Marconi's wireless telegraph. The first voice to pass over radio waves and be heard by another person was that of Canadian professor Aubrey Fessenden. A year before Marconi's historic telegraph signal across the Atlantic, Fessenden used a spark-gap transmitter to speak to an assistant a mile away. This is what he said:

"One, two, three, four, is it snowing where you are, Mr. Thiessen? If it is, would you telegraph back to me?"

We don't know how Thiessen responded, but it's clear that Fessenden belonged to the Alexander Graham Bell School of Less-than-memorable First Messages.

Lee De Forest, inventor of the triode tube, made the first radio broadcast to the public from the Metropolitan Opera in New York in January 1910, but few heard it because only a few people had receivers. Not until after World War I would commercial radio broadcasting begin to become common.

The same was true for two-way communications by radio. Amateurs started broadcasting and receiving voice messages in the 1920s. A police radio system also was created but wouldn't be fully realized until the 1950s.

The mobile phone was a natural commercial extension of two-way radio. But in 1934, the Federal Communications Commission (FCC) was created, and part of its job was to manage the spectrum of radio frequencies to prevent transmission chaos. Different frequencies were allotted for different purposes. The FCC gave priority to creating emergency systems and allocated no frequency bands for commercial two-way radio until after World War II.

The first mobile phone system was put into service in St. Louis in

1946. These phones were installed in vehicles. Incoming calls were transmitted from a single tower antenna connected to ground lines, but five receiver antennas were placed around the city to pick up outgoing calls and direct them to the central antenna. Only one-way conversation was possible at a given moment. A user had to push a button to talk, then release it to listen.

Soon mobile phone systems appeared in other cities, but the service was always expensive, limited in range, and susceptible to interference. The number of people who could use it was restricted, and the circuits always were crowded because of the scarcity of frequencies allotted, causing calls to be difficult to make or receive. Thirty years after introducing mobile phones, the Bell System had only 44,000 subscribers with another 20,000 on five- to ten-year waiting lists.

Not until the late 1980s would radio telephone service begin to become widely available through the cellular phone.

But well before then came another technological breakthrough that eventually would make modern cell phones possible: the integrated circuit.

The IC, as it is known in the trade, was invented in 1958 by two people working independently in separate companies. One was Jack Kilby, a recently hired engineer at Texas Instruments in Dallas, Texas, and the other was Robert Noyce, one of the Shockley deserters who started Fairchild Semiconductor in Palo Alto. The idea of the IC was to put an entire circuit—transistors; capacitors, which condense and hold electric charges; resistors, which direct current; and connecting wires— all into a single tiny semiconductor chip.

The original ICs, one in silicon, the other in germanium, contained only a single transistor, three resistors, and a capacitor, but ICs would quickly grow more complex. Now a single minuscule chip can have 125 million transistors, or more.

ICs, like the transistor before them, were less expensive, more efficient, and produced higher quality than the parts they replaced. So tiny were they—about the size of a pencil point—that electronic devices could be made far smaller. Used first in military applications, ICs quickly became the basis of all computers.

By the beginning of the twenty-first century, these minuscule chips would be a worldwide business bringing in trillions of dollars, and RF Micro Devices would be a prominent part of it.

The cell phone business developed slower than the computer revolution for several reasons. The FCC was reluctant to release more frequencies. A huge infrastructure had to be built to make it possible. And the technology that would allow it to work was not yet available.

In 1968, in response to an offer from the FCC to free more frequencies for mobile phones, Bell Labs created a system that divided a service area into cells of hexagonal grids of about 10 square miles, each with low-power tower antennas to send and receive calls.

Each cell cluster would have a single switching station to hand off calls as a subscriber moved from cell to cell, or from one cluster to another. Because of the low power of transmissions, a cell only one cell removed from another could use the same frequencies without interference, thus greatly increasing the number of calls that could be made within a cluster.

The first completely portable cell phone was built by a team at the Motorola Corporation headed by an executive named Martin Cooper. A first ceremonial call was, of course, required, and Cooper made his to a competitor, Joel Engel, the chief of research at Bell Labs. We only can guess that his message went something like this: "Nah-nan-aa-nan-naah."

In 1978, AT&T set up the first test cellular phone system in Chicago with 2,000 subscribers. Three years later, Motorola, in partnership with American Radio-Telephone, started their own trial system in Washington and Baltimore. But not until 1982 did the FCC authorize commercial cellular service.

Cell phone use was restricted to major cities in its beginning years. Early cell phones were so cumbersome and heavy—primarily because they required such huge batteries—that they had to be lugged around in small suitcases. They were called bag phones, and users often had to fumble to get to the receiver if the phone happened to ring.

Despite that, demand for cell service was so great that by 1987 more than a million subscribers had signed up, outstripping the capacity of the available systems as well as the prevailing transmission technology.

In an attempt to meet the increasing demand for service, in 1987, the FCC set aside a higher frequency band for cellular service that would require new transmission technology and new phones to use it. The following year, the Cellular Technology Industry Association (CTIA) was organized to coordinate the efforts to meet these new technology demands.

That was the same year that Bill began traveling to cell phone makers to find out what their needs were and to begin designing products to meet them.

Early cell phones operated on analog service, similar to FM radio, which modulated the voice onto signals that were transmitted and received on the same frequency, limiting the number of channels that could be used.

It was clear, however, that the future of wireless devices, such as cell phones, lay in digital technology, using the code of zeroes and ones that operates computers. In digital cellular service, the voice signal would be converted to this binary code and condensed, allowing far greater numbers of calls to be transmitted over the same frequencies.

The new transmission technology would have to provide for both analog and digital services, and the CTIA wanted the equipment in place and new, more sophisticated phones ready for testing by 1991— coincidentally the year that Bill and Powell started RF Micro Devices.

Our former employer, Analog Devices, was at the forefront of making digital signal processing possible, and knowing that this technology was coming was part of the reason Bill had tried to lead the company into making radio frequency integrated circuits. The company's decision to pass on it was our good fortune, for we now were far ahead of others in designing products for this new and bountiful market.

That was why I was so confident about being able to compete with giants such as Motorola and AT&T when I was showing off our first chips to my dad in our company's small quarters during the summer of 1991.

Of the seven chips that Bill and Powell had produced in small but costly prototype batches, all but two, both quadrature modulators, were receiver parts.

Four of these chips were based in silicon—two low-frequency high-gain amplifiers, a quadrature demodulator, and a quadrature modulator—but all were in limbo, awaiting the money that would allow us to produce and test a second, redesigned batch.

The other three chips—two low-noise amplifier/mixers and a quadrature modulator—were in a newer medium, gallium arsenide, a material made of the elements gallium and arsenic, a far better conductor than silicon. Bill had designed these at the instigation of Analog Devices president Ray Stata, because Stata wanted to get some use out of his investment in Gigabit Logic, a company that used this process. But Gigabit Logic had been shut down after producing our prototype chips, and we had been forced to search for another foundry with the technology to supply them. Only these three chips were anywhere near being ready for market, however, and not even they were functioning fully as designed. They worked. They just didn't work well enough to meet all the specifications that potential customers surely would require.

If you will forgive an analogy from a stock car racing fan, it was like a racing team building a 600-horsepower engine and discovering that it put out only 400 horsepower. They still could put it in a car and take it on the track, but if they didn't want to be humiliated, they had to go back into the engine, find out what was wrong, and fix it.

This situation was not unusual in the electronics trade. Bill likes to point out that engineers spend most of their time fixing what they set out to create. Only a fool, he says, would think that an engineer could sit down at his computer, design a chip, send it off for layout and manufacturing, and have it come out working perfectly on the first try.

If that happens, it's the next thing to a miracle. The real work comes in the fixing. That's why engineers have to be unflappable, optimistic, and especially persistent. And those who work with them have to share those qualities as well.

Sometimes, this can cause problems in a company, as Bill also likes to point out.

The manufacturing folks will be screaming, "How can we make something that works when junk is all we get from engineering?"

The marketing people will be pitching tantrums because they have customers with money in hand waiting for a product that they expect to work.

Bill knew that such a situation never would occur with Powell and me. We had been together a long time. We understood the procedure and each other. We expected problems, never allowed them to unsettle us, and did whatever was necessary to overcome them.

This attitude went back a long way with me. My dad was a supervisor at a company that made office furniture. He was a highly methodical man who was always solving problems, technical and otherwise. There were two kinds of people in the world, he used to say, problem makers and problem solvers, and it was far better to be the latter. For almost as far back as I can remember, he impressed on me that a person who could solve problems would always be in demand, always be welcomed, and always be able to make a living.

To him, a problem was never to be viewed as anything but a challenge to be mastered. His example and advice proved to be a great advantage in my sales career, and I'll never forget how complimented I felt when early in our relationship, Bill told me, "Jerry, the thing I like about you is that you always run toward problems, never away from them."

Heaven knows we had plenty of problems to run toward in that first year of the company's existence. When I look back now, I'm astounded by how much we actually accomplished in spite of all the time and energy we were forced to devote to chasing funding.

In addition to the seven chips for which we already had samples and prototypes, we had another in layout, and Bill was working on the designs of several others.

Moreover, despite all of our problems, we even managed to land our first two customers late that year, although neither wanted the products we already had developed.

One of the reps I had lined up to cover Canada called in the fall to say he had a client who was going through an ordeal trying to get two parts made.

The client was Digital Security Controls in Concord, Ontario. It was a start-up company, too, just getting going. Its founder and president was John Peterson. He wanted to manufacture wireless motion and smoke detectors and needed transmitter and receiver chips for them.

Bill and I spent much of a day bouncing around the sky in small commuter planes to get to Canada to meet with John and the engineer in charge of this project, Michael Yau.

John and Michael had been working with another company for two years to get these parts made with no real success. They told us about all their problems, showed us what they needed, and what hadn't worked.

Bill decided immediately that we could design chips to fit their needs. But their previous experience had left them a little gun-shy, and they weren't certain. Bill and I had to make several more bouncy flights to Canada on my frequent flyer miles before we persuaded them that we could do the job.

This was a custom order for which we would be paid fees at certain stages of development for non-recurring engineering costs. When the chips were perfected, we would get them manufactured and begin selling them to Digital Security.

Our second customer was a Japanese company called Nippondenso, later just Denso, with a plant in Long Beach, California. Once part of the Toyota Motor Company, Nippondenso manufactured automotive parts, air conditioning systems, and other products. It was planning to build mobile phones, and that was how we came to Nippondenso's attention.

Nippondenso wanted a power amplifier in the 900 megahertz range for a phone that Toyota wanted to install in its luxury vehicles. A power amplifier takes a weak current and makes it stronger. In a mobile phone, it would take power from a battery and boost it enough to transmit the signal to the receiving tower. No power amplifier ever had been designed as an integrated circuit, but Bill had been working on a plan for one for nearly a year, and he flew to Long Beach to talk with the engineers about their needs. He was convinced that we could do it in a technology that previously had been used only for highly sophisticated military and space equipment.

In December, we prepared a formal proposal and got a verbal

acceptance even before we knew for certain that we would have the funding to keep the company going.

Nippondenso agreed to pay us $200,000 in engineering fees over the coming six months, and we intended to have the power amp in production late in 1992. We expected revenue of $750,000 a year from it over the next five years.

Our contract with Digital Security Controls called for us to get $185,000 in engineering fees for the two chips we already were designing for them, which we also planned to have in production by the end of 1992. We anticipated revenue of $1 million a year from those for the indefinite future.

This, along with the interest we were getting in the products we already had designed, caused us to project that we would have revenue of more than $1 million in 1992. We expected sales of more than $5 million in 1993. We hoped to double that in 1994, and double it again the following year.

But despite all of our bold confidence in our ability to project so rosy a future, we somehow weren't able to see the obstacles that lay ahead, or the troubles that soon would befall us.

4
The Mysterious Disappearing Chips

To understand the situation RF Micro Devices was in at the end of our first year, it's necessary to know that the technology that most intrigued Bill never had been used in a mass-market product.

The wonder is why.

Gallium arsenide, the semiconductor Bill had chosen for half of our original chips, conducts electricity six times faster than silicon, the material used then in most integrated circuits. That means it can do more work faster, an honorable goal for any marketable device, not to mention the dream of all employers for the people they hire.

These first chips Bill designed in gallium arsenide were done in a technology called MESFET, which stands for metal semiconductor field effect transistor.

For those of you who are electronically challenged and are just beginning to be perplexed by these technical terms, give me a little more space and I may be able to deliver you into a state of total confusion. But I promise that I'll be trying hard to do otherwise. And I won't dwell on this any longer than it takes to explain the situation.

MESFET is not the only technology for which gallium arsenide can be used. Another is quite a mouthful as well—heterojunction bipolar transistor, or HBT—which has nothing to do with the condition that causes people to go from extreme highs to extreme lows, and vice versa, although it eventually would deliver us to some.

The heterojunction bipolar transistor was patented in 1948 by none other than William Shockley, who headed up the Bell Labs team that invented the transistor. But it was not until more than two decades later that the technology was developed at government expense to use it in gallium arsenide. Even then it was employed only in the most sophisticated military and space equipment, and at costs so forbidding that nobody thought it ever would have commercial possibilities.

Many engineers simply didn't trust gallium arsenide because it had an inherent flaw. Under certain conditions, the two elements would fuse together and cause the chip to cease to function. Although Bill was wary about the technology, he had studied the gallium arsenide processes, knew that chips using them would be highly reliable, and thought they could be made inexpensively enough for consumer products.

The HBT process in gallium arsenide had several advantages over MESFET, and that was why Bill had chosen it for the power amplifier we had agreed to produce for Nippondenso.

Because HBT needed only a single power source, whereas MESFET required both positive and negative supplies, the HBT chip demanded far less power from the battery. This would greatly increase talk time on a cell phone, for example. The more complex design of HBT chips also reduced distortion and provided a stronger and clearer signal. Better quality and greater usage, like people who solve problems, always would be in demand.

Another big advantage for HBT was that MESFET chips had to be designed horizontally, but HBT chips were vertical in structure. This allowed for smaller circuits and dramatically reduced the chip size.

Semiconductor chips are manufactured in round wafers of varying diameters, each containing thousands of chips. Far more HBT chips could be included in a single wafer than MESFET chips, dramatically cutting manufacturing costs and the eventual price of the chips. These smaller chips also had great appeal because the devices in which they were used could be made slighter in size and lighter in weight.

Only a handful of foundries were capable of producing gallium arsenide chips, and fewer still could handle the HBT process, which had been simultaneously developed with varying success by only a few companies with U.S. Department of Defense funds. One of those companies, Texas Instruments, had abandoned it by this time, but later would return to it.

Bill had struck up a relationship with another of those companies, Rockwell International. Rockwell not only was a major Defense Department contractor but it had built the lunar lander Apollo. Bill turned to Rockwell, which had a gallium arsenide foundry in Newbury Park, California, after the demise of Gigabit Logic, the company which originally supplied our gallium arsenide chips. Some people Bill knew at Gigabit Logic had gone to work at Rockwell, which was only a short distance away. We planned to do all of our gallium arsenide chips there as soon as we got the investment money to begin production.

The folks we were dealing with at Rockwell assured us that we would have no problems getting the power amp for Nippondenso done in HBT. They would work with us, they kept telling us, to do whatever we needed to get the chip made at a reasonable cost, which, we had told them, would be in the range of $2.50 or $3 at maximum. The problem was that we couldn't get a price out of them. We kept pushing and pushing, and they kept putting us off.

We were stunned when we finally got their quote. They wanted $13.25 per chip. This meant that we would have to sell the chip to Nippondenso at twice that to make a profit. We had quoted the part to Nippondenso, tested and made ready to plug into its circuit boards—packaged, as we call that procedure—at $14.50 in the quantities Nippondenso said it eventually would be needing. At that price, we'd be losing whopping amounts on every shipment.

We knew that these chips could be made for far less than $2.50 and that Rockwell still would be making a good profit on them.

Bill flew to California to try to negotiate the price down to a reasonable range. But nothing would budge them.

He returned utterly disgusted and so pale and weak that he looked as if he couldn't take another step. Not only did the Rockwell folks remain intransigent, he told Powell and me, but they took him out for a lunch of spicy Indian food that made him deathly ill. The afternoon negotiations were frequently interrupted by his frantic dashes to the men's room. He spent most of his redeye flight back sprinting the jet's aisle. He felt as if he'd run all the way home, he said.

Bill thought that Rockwell had deliberately misled us all along—not to mention giving him food poisoning—and he wanted nothing else to do with the company. We already had entered agreements with Rockwell for two of our other gallium arsenide chips, also at a price too high, as it would turn out. That was a deal we were stuck with for the time being, though, and freeing ourselves from it eventually would become a nasty affair. Our feeling was that Rockwell had such a military mind-set that it was incapable of functioning in the real world, where price mattered.

We had to find another place to get our power amp chip made, and we had no idea whether we would be able to do it at a price that would allow us to keep our commitment to Nippondenso.

As it happened, I had seen an ad in a trade magazine only a week or so earlier in which another major defense contractor was, for the first time, offering foundry services for small chip runs, including gallium arsenide HBT. The company was TRW in Redondo Beach, California. I dug out the ad and dialed the number without any hint that this would turn out to be one of the most important phone calls in our company's history—and no small matter for TRW either.

The person I talked with was Scott Andrews, a marketing director. I explained our situation and told him I'd like to come out and discuss the possibilities of TRW producing a power amp chip for us. He said he'd try to set up a meeting and call me back.

In the meantime, we learned that TRW's gallium arsenide HBT process was different from Rockwell's. It would give us a smaller chip with better quality. If we could get it at a reasonable price, we really would be on to something.

A few days later, Scott called to say that the meeting was set, and I found myself on my way to California.

That meeting turned into an all-day affair, with several TRW officials involved, including Scott, Barry Dunbridge, Aaron Oki, and others dropping in and out.

Looking back, I only can wonder what they thought, but I can guess that they must have seen me as either awfully cheeky, or maybe out of my mind. Here I was trying to explain to executives of this huge corporation how three guys with no money (although I didn't tell them about that) in Greensboro, North Carolina, not only planned to penetrate the quickly emerging radio frequency market but fully intended to dominate it. I certainly was doing my dead-level best to make them think that we were granting them a favor by including them, and as events would turn out, that would prove to be the case.

At the time, TRW wasn't even doing production work for other companies at its foundry, and here was this mild-mannered, and in reality far-from-brazen character—me—proposing not only that TRW change its course of business by taking on this power amp for us, but do it at a price so low that I was worried after our Rockwell experience that they might laugh in my face.

We needed these chips at a cost of about a dollar to $1.50, I told them, setting a price low enough to negotiate, but we would be ordering in such large quantities that it still would be lucrative for TRW at that price. Remarkably, nobody even hinted that that might not be possible.

Still, I didn't know what to think after this meeting. I heard what I expected. They would take all of this under consideration and get back to us.

It was with more than a little anxiety that we awaited their decision. If they said no, we really had nowhere else to turn and would have to tell Nippondenso that we couldn't fill the order.

Bill was nearly finished with the design of the chip and was confident that it would work and make a lot of money for us, but we were left hanging in suspense about what was happening in Redondo Beach.

Fortunately, TRW didn't make us agonize very long. Little more than a week later, I got a call from Scott saying they had decided to take on our project. This was great news, of course, and it marked the beginning of a long and profitable relationship between TRW and RF Micro Devices.

Now we could draw up a contract with Nippondenso to add to the one we already had signed with Digital Security Controls.

By the time we had our first board of directors meeting on March 10, 1992, only days after getting our first installment of investment money, we were able to show off these signed contracts and to report that we were well on our way to developing test-run prototypes for our first cash-producing chips.

That board meeting was held in the conference room at the front of our little gray module, the only room with a window. The window stretched from wall to wall and offered a great view of the narrow parking lot and the traffic on Friendly Avenue when the blinds weren't drawn. It also allowed anybody approaching the front door to see whatever was going on inside, which was never anything likely to attract an audience.

The room was so small that it barely could accommodate the table and eight chairs for the seven board members, and for Dean, our comptroller, who had to sit in to present our dismal figures.

Bill had a great fondness for palm trees, and he thought that a nice palm would be just the thing to brighten up this otherwise drab room, the only place in the building with enough natural light to allow a plant to thrive.

He lugged in a big one. I'm uncertain what kind it was, not being an authority on matters of flora, but it was a sure-enough palm tree. It gave us a little sense of the tropics, and it was far more visually appealing than the view from our window.

That palm took up a lot of room, and somehow at that first meeting Walter Wilkinson, our most supportive investor, ended up sitting beneath it after managing to fight his way into the chair without benefit of a machete.

As the meeting began, I looked over and saw Walter, a Harvard man, sitting in total earnestness with a palm frond draped over his forehead, and it took all of the self-control I could muster to stifle a laugh.

As the meeting wore on into ever deeper earnestness, I would glance over at Walter now and then and occasionally catch him casting a wary eye upward as if he might be expecting a monkey to start hurling coconuts down to disrupt the proceedings. At board meetings to come, there would be times when I would wish that we did have such a monkey, and that I could select his targets, but I won't get any more specific than that.

The board was made up of our investors, Walter, Al Paladino, Erik van der Kaay, Ray Rund, plus Bill, Powell and myself. One of our first orders of business was to elect a chairman.

Since we were outvoted on the board, we thought Bill should hold that position, and the decision was unanimous. Bill seized on his newly sanctioned authority to suggest that we increase the board to eight members. We wanted to add an old friend of ours from Analog Devices, a genuine character named Hank Krabbe, who now was retired in California, to even up the vote. Needless to say, there was little enthusiasm for this from the majority, although they were polite in nudging the notion toward Neverland.

The primary concern of the investors at this meeting was the same as ours: to get our original chips ready for market, to hire the technicians and engineers needed to help us get that done, to determine which foundries we would use to manufacture those chips, and to begin lining up customers to buy them.

We, of course, had been moving on those goals all along. We were fortunate in one respect. We could have been without a sales force at this point.

Burr-Brown's offer to finance us had included a stipulation that our products be sold through its in-house sales team. By the time we got the Burr-Brown offer, I already had lined up sales reps to cover almost all of the United States and Canada, and much of Europe and Asia, and I wasn't looking forward to having to call all of these folks and say, "I'm sorry, but we're not going to be using you after all."

Luckily, I had put off those unpleasant calls until the final contracts with Burr-Brown were signed. That saved me from having to rebuild an entire sales force after the deal fell through. Since then, I had filled the remaining gaps in the sales team.

Beginning to build our technical team was not that difficult either, because we already had people in mind for several positions. Jill Howard had applied for a technician's job long before we had money to hire anybody. She lived in Greensboro; had been driving to work in Durham, 50 miles away; and was eager to be rid of the long commute. She began working with Powell on March 23.

A week later, we hired an engineer, Bob Hicks, to be in charge of testing chips. All of us had worked with Bob at Analog Devices. He later would marry Powell's daughter, Wendy.

We also had two other engineering prospects lined up as soon as they could break free of their current situations. One was Kellie Chong, who worked for another small high-tech company started by one of our friends from Analog Devices. She would join us in May and begin working on the Digital Security chips. Bill's son, William H. Pratt, who's called Billy, would come to work for us in June, after getting his engineering degree at Georgia Tech.

• • •

As we added people, we became more cramped in our small quarters. Bill was deeply intent when he was working hard, and that was all the time during this period. When he was in that state, he also smoked cigarettes with a fury. He would open his office door and a cloud of smoke would boil out that rivaled the fog that regularly enshrouds San Francisco.

A couple of employees who worked near Bill's office—and I won't get into names here—came to me at one point to complain that they didn't think they could stand the smoke any longer.

At times they barely could breathe, they said. Their clothes stank. Their sinuses were in a constant uproar. They couldn't even sleep at night, so rapidly did their blood course with nicotine. Something had to be done!

And I was the only person, they were convinced, who could talk to Bill about it.

I wasn't about to talk to Bill about it, of course. If Bill couldn't smoke, we might never have any chips to sell.

Instead, I concocted a secret plan.

Although Bill, Powell, and I continued to work until almost midnight every night, long after our employees were gone for the day, we never worked on Saturday and Sunday.

One Saturday morning, I had a crew from a heating and air conditioning company meet me at the office. Over that weekend, they installed a high-powered exhaust fan with a vent directly over Bill's desk that fed into a duct leading to the parking lot at the back of the building. The fan was wired into the light switch.

When Bill opened his office door and turned on the light Monday morning, the fan came on. If he'd been wearing a toupee, it would have snatched it right off his head. Even the framed pictures on the walls shimmied as if they were about to take flight through the vent. I'm exaggerating here, I admit, but not a lot.

When I'd go in to talk to Bill, his hair would be standing on end, and his tie would be fluttering up from his shirt, as if it were trying to lick him in the face. But Bill never noticed.

It wasn't until a few weeks later, when he was going through the bills to be paid, that he came to me and said, "Do you know anything about an exhaust fan?"

My Quaker upbringing always required me to fess up immediately, and I did. Bill wasn't happy about an expenditure that he considered to be so frivolous, and he harrumphed about it for a while, but the people who worked around him smiled with new contentment.

Late April brought us one of the great moments in the history of

RF Micro Devices. An envelope from California arrived in the morning mail. Inside it was a check from Nippondenso for $44,000, the first payment on engineering fees for the power amplifier we were developing for the company.

Fourteen months after the company was founded, we had earned our first revenue. All the out-go on our ledger now would have a bit of ballast on the other side.

We all gathered to pay homage to this exquisite moment. We passed the check from hand to hand, and held it out at arm's length to admire it. A few of us could have kissed it; I'm not sure. It was a beautiful thing, pale gray with "Nippondenso" and the check number—144938—printed in bright red.

Ignoring the expense, we even took it to Kinko's to get color copies made before we deposited it so that we could display it, the way a diner owner might put the first dollar from a customer in a frame and hang it behind the cash register. Indeed, one of those copies hangs in a glass case in our headquarters lobby today.

By mid-May we had in hand batches of six of our original seven chips, and we were busy testing them to see if they needed any tweaking before I began sending samples to potential customers, some of whom already were requesting them. By this time, too, Powell was finishing the layout of our power amp for Nippondenso, and we soon would send it off to TRW for the first chip run.

Things were going well, and we were feeling good about ourselves, even though experience should have taught us by now that this was likely an omen warning us to brace ourselves for calamity.

Trouble hit like a hammer blow when we received our first batch of Nippondenso chips in July. These chips were what we termed a hybrid, meaning that they had been mounted in a base, this one ceramic, with other elements vital to the chip's function in the mobile phone for which it was designed.

Bob Hicks took these chips straight to the lab, hooked one to the test board on the stage of a microscope, and turned on the juice. The digital gauges on the power meter and spectrum analyzer showed that this tiny amplifier was putting out an impressive amount of power.

Then the gauges suddenly went blank.

Bob turned to the microscope to see what had happened and raised his head with a puzzled look on his face.

"What's the matter?" Powell asked.

"Well, it's gone."

"What do you mean, gone?"

"It's not there," Bob said. "Take a look."

Powell looked and shook his head. The chip had disappeared, leaving a nearly invisible molten puddle in its place.

Powell came to tell Bill and me what had happened. We all trekked back to the lab, and as we watched, Bob hooked up another chip, turned up the power, and pressed his eye to the microscope as we watched the gauges. The chip was showing great performance until the gauges blanked again.

"It just went *pfffft!*," Bob said, "and it's gone."

We all took turns at the microscope watching these chips go *pfffft!* With every one we could see dollars—and possibly our company's future—going up in invisible puffs of smoke.

It was clear that as tiny as these incinerations were, this was no small problem. It could be a major setback. Bill had no idea what was happening in these chips, and neither did the rest of us. We didn't know whether it was a mistake in our design, a problem with TRW's process, or the inherent flaw in gallium arsenide that caused so many engineers to mistrust it.

We realized this likely was going to be no quick fix, and Nippondenso was expecting us to begin delivery of these chips by November. We had to let Nippondenso know that a problem had developed, and within days the company dispatched an engineer to determine how serious it was.

Don Green was a sharp, no-nonsense guy who once had been a scientist with Bell Labs. We sat him down at the microscope so that he could examine one of the chips himself.

"They look good when you see them," Bill told him.

"I don't see any problems," Don said, peering into the eyepiece.

"Just keep looking," Powell said.

Suddenly, Don looked up with a start. "There's no chip!"

"That's the problem," Bill said drolly.

We spent hours in discussion with Don, theorizing about what might be causing this. We had alerted the engineers at TRW, and they, too, were looking into it, we assured him. This was a new concept in an exotic technology, and difficulties had to be expected. We would figure out what was going wrong and fix it. We just didn't know how long that might take.

Don carried a micro-recorder in his pocket into which he spoke notes to himself. During the breaks in our meetings, he would go outside to the parking lot and pace about intently delivering long soliloquies into his little recorder. I hid behind the palm tree in the conference room and peeked out at him to try to get a reading on what he might be thinking. He certainly didn't look happy.

I could just imagine what he was telling that recorder to relay to his bosses: "Not only can these guys not make a chip that works, they can't even make one that you can see for very long."

I dreaded having to let our VCs know about this problem, but there was no way to avoid it. Perturbed might be a mild word for their concern. For people who put up large sums of money, we were beginning to learn, catastrophe lurks in every snag; disaster looms over every setback. At our scheduled board meeting on August 4, they demanded to know exactly what we intended to do about this mysterious disappearing chip.

"Well," Bill told them without elaboration, "we're going to fix it."

Powell and I couldn't help but smile—inwardly, at least. Our long experience with Bill told us that was exactly what would take place. We didn't know exactly how. We didn't know exactly when. But we had no doubt that this chip would be fixed.

Although our investors were aware of Bill's expertise and skill, they hadn't yet experienced his strong will and determination when it came to overcoming problems, thus could be temporarily forgiven their lack of forbearance.

Nippondenso wasn't aware of Bill's better qualities either. All it wanted was a chip that worked at the time it needed it, and although we were on the phone to Nippondenso's executives daily keeping them informed, they were not appeased. Don Green, whom we liked a lot, returned for more discussions and more pacing in the parking lot talking intently into his recorder. He made his company's displeasure clear, and warned us that it was losing patience.

We were not surprised when a short time later we received formal notification from Nippondenso that it was cancelling the contract for failure to meet the production schedule. It was a low time for all of us, but that didn't deter Bill. He kept right on working to fix that chip.

From the beginning, our investors had insisted that we hire an outside chief executive officer. We didn't like that. Powell and I thought that Bill should be CEO. We couldn't imagine anybody else in the job. His vision and skill had created the company, and we thought that he should lead it.

Bill didn't think so much of bringing in an outsider either. He was willing to do the job—and perfectly capable. His years of guiding the Computer Labs Division of Analog Devices through its tremendous growth had proven that.

Our investors had made an outside CEO a stipulation of funding, however, and wanted to start an immediate search to fill the position. We attempted to delay it—and for good reason.

Our immediate challenge after receiving the first of the investment money was to get our chips ready for market. We needed to be hiring engineers and technicians to accomplish that, not a CEO. A CEO search would be a distraction to the real work that needed to be done. Also a CEO's salary would be a drain on our initial funds that could be put to better use. We thought that we should wait at least a year before beginning a CEO hunt.

But at the first board meeting on March 10, Walter suggested a candidate for the job. At the next meeting, on April 21, the subject came up again. Walter offered a second candidate. Al had one as well, and Ray another.

We realized that it would be futile to offer a candidate of our own, but we had gotten the VCs to agree that nobody would be hired for the job who didn't meet our approval.

If we had to have a boss, we wanted it to be somebody who wouldn't act like a boss, somebody who not only knew the radio frequency business inside and out, but who also respected us and was as enthusiastic as we were about building this company.

We also wanted somebody caring, congenial, with a sense of humor. In other words, we wanted somebody we could get along with and, most importantly, somebody who would share our vision for the kind of company we wanted RF Micro to be. We wanted somebody who would help us build an exciting, creative, rewarding, and harmonious atmosphere for all who worked with us.

It was especially important to us that we have an open environment in which employees considered themselves part of a family and felt free to speak their minds, make suggestions, and attempt new things without worrying that they would be criticized or reproached if they didn't work out. We wanted people to have the confidence to make decisions on their own and to take responsibility without dread of failure.

I've always believed that the most innovative and successful people are those who aren't afraid to fail. To bring up my passion for racing again, the best drivers are always those who aren't scared of wrecking. They push themselves constantly to the edge, never back off unless they are forced to, and are never reluctant to take prudent chances.

Bill, Powell, and I had always believed that the best people in our business were the risk takers, too. And we wanted people working with us to know that they could try new things without worrying that somebody would be hovering over their shoulders eager to find fault and to blame them if they spun out, or hit the wall.

Most of all, we wanted the people who worked with us to have as much fun as we were having and to look forward to coming to work every day because they knew it would provide the same pleasure and fulfillment that we were feeling.

Over a period of a few months, the three of us had interviewed a couple of the CEO candidates suggested by the investors, but neither turned out to be a proper fit.

Then Erik van der Kaay suggested David Norbury, with whom he had worked at a Silicon Valley company called Avantek several years earlier. I stopped by to meet Dave on a sales trip to San Jose and thought he would be a good match. We weren't the only company interested in Dave, however. He had an interview in Boston and decided to make a side trip to North Carolina with his wife, Barbara, and two children, Michael and Samantha, as a vacation, with a stop to see us.

Dave had a master's degree in electrical engineering from Stanford and a master's in business administration from the University of Santa Clara. He recently had sold a high-tech company that he had started two years earlier in Silicon Valley, Polylithics. Ray Rund had been one of the VCs who had funded the company, and he was especially impressed with Dave.

Later, Dave would say that when Erik called to ask if he would be willing to take a look at our little company, he thought: *Greensboro, North Carolina? Are they nuts?* He couldn't believe that somebody would start a company to make radio frequency semiconductor chips so far from Silicon Valley.

I was a couple of years younger than Bill and Powell, and Dave was about a decade younger. It was clear, though, that he knew the RF business. Beyond that, he was affable, easy-going, and had a keen sense of humor. We all thought that we could get along well with Dave. I was especially impressed that he was a ham radio operator, like myself, and that his major hobby was woodworking, one of my dad's great passions.

Dave and his family toured Greensboro and the surrounding area while they were visiting and decided that they liked it. After quizzing us about our hopes for the company and studying the business plan, Dave agreed with Erik that we were well ahead of others and on our way to something big. And he wanted to be part of it.

With our agreement, the board voted at the August meeting to hire Dave, and he joined us on September 9 as the fifteenth employee of RF Micro Devices, although two earlier employees had left by this time.

I was particularly happy to have Dave on board. Up until that point, I was the one who had to handle all of the day-to-day dealings with our investors. Now all of that would fall to Dave, and I could stash away my coconuts, return that monkey I was training to the pet shop, and devote myself to letting people know what a formidable force RF Micro Devices intended to become in wireless communications.

Unfortunately, we were getting off to a more sluggish start toward that ambitious goal than we had anticipated. We had lost one big customer,

Nippondenso. And we wouldn't be able to begin shipping chips to the first customer we signed up, Digital Security Controls, by the fall deadline either. This was in part because of problems we'd had with the design of the chips, but also because Digital Security kept discovering more things that the chips needed to do and changing the specifications.

Because of these problems, we would end up with far less revenue than the $1 million we had forecast for 1992. We would take in just $212,000, only $22,000 of that in sales, mostly for sample parts, the rest in engineering fees from Nippondenso and Digital Security Controls.

Nonetheless, most of our original chips were now ready for market, and we had begun getting respectable orders for several of them. We had freed ourselves from Rockwell and found another foundry to do our MESFET gallium arsenide chips.

We also had continued adding people until it was hard to turn around in our tiny quarters. By the end of 1992, our staff would number 16, more than five times the number of a year earlier, including engineers Jeffrey Scalf, Alan Nicol, Leonard Reynolds, David Adams, and a technician, Clarence Oliver.

Despite our difficulties and disappointments, we were well on the way to working out the problems with the Digital Security Controls chips. And despite the loss of our Nippondenso contract, Bill had worked doggedly on to fix our power amp chip, which we now had dubbed Son of Nippondenso.

By October Bill had figured out the problem. Transistors in the chip were hogging the power, causing overheating and meltdown. Bill added emitters to keep the power flowing, and in November, we sent the redesigned chip off to TRW for a prototype run.

Those chips were due back before Christmas, and we were hoping that they would bring us a very merry one. But Christmas for my family would be tainted with sorrow.

My father-in-law, Gene Stewart, died on December 21, and I wasn't at work when the Son of Nippondenso chips arrived.

The following evening, I was with my wife, Linda, and her family, receiving guests at Sechrest Funeral Home in Archdale in northern Randolph County, adjoining High Point. Bill and Powell came together from work to pay their condolences. Afterward, they pulled me aside to tell me that the new chips had come in and were working perfectly.

Nobody had to tell the three of us the implications of this. We had something that nobody else had, the first and only high-quality, integrated-circuit power amplifier—and every cell phone and cordless phone made from this point would need it, or one similar to it.

We knew this chip would do great things for our company. We just had to get the word out about it. And that job was mine.

5
Where's King Kong When You Really Need Him?

The official name of the chip we had dubbed Son of Nippondenso was the RF2103, or 2103 for short. It was given that dry, nondescript handle because—well, because that's the way engineers do things. All other producers of semiconductors used similar numbers, so we did, too.

The original numbers we picked for our products had no particular meaning. Bill just snatched them out of the blue.

Once established, though, the numbers assumed significance. The first two digits indicated the type of chip that it was. Power amps were 21, modulators 24, demodulators 27, and so on. The second two digits indicated that particular chip's position in the product series.

For example, the first power amp, the 2101, was the one Bill was designing when we got the custom job for the failed Nippondenso chip, which became the 2102, although neither of those chips actually ended up in production. I should point out, however, that this system was subject to change for reasons both logical and not.

One thing was certain: the 2103 was a powerhouse of a chip, and we had no doubt that it was going to be a big success. We even thought it could be a company maker, the way the first memory chip—the 1103 created by the Intel Corporation, which was co-founded by Bob Noyce, the co-inventor of the integrated circuit—turned that company into the world's largest manufacturer of semiconductors.

But the 2103 was slower in taking off than we anticipated. Both we and TRW sent out news releases about it to general media outlets early in the year. We got little notice from these efforts, however.

Our greater hope lay in the electronics trade magazines, which reach the people who make the devices that could use our parts. Those magazines allow free space for new product announcements, and we sent out photos of the 2103 and specification sheets to all of them.

I admit that there is a certain tedious sameness to photos of semiconductor chips. They don't exactly have the photogenic appeal of a *Sports Illustrated* swimsuit-edition model, but engineers love to look at them anyway.

Our hot little number—the 2103, clad in midnight black—miraculously made the cover of *Microwaves and RF Magazine*, which eventually would name it their product of the year. They even would present us a trophy for it.

These announcements brought an exciting amount of requests for additional information and samples, but we didn't fully grasp the reality of what we were facing until March 18, 1993, when I presented a technical paper that Bill and I had written about the 2103 at the RF Expo West Conference smack in the middle of Silicon Valley at the San Jose Convention Center.

The room was packed with engineers and others in the trade, and when I finished my presentation and asked for questions, I found myself facing not just open skepticism but downright hostility.

It quickly became clear that many people in the room simply didn't believe that gallium arsenide HBT technology could be reliable at a reasonable cost. In their minds, it still was a flawed technology that only the government could afford to make work. They just weren't willing to believe that it was practical for an integrated-circuit power amplifier, for which it never had been used.

"Look," I told them, "you don't have to believe me. I understand your skepticism. But write down this number—910-855-8085. Call and ask for a sample. Forget what I've said. Just put it in your lab and test it for yourself. Then we'll talk."

The unexpected reaction I got in San Jose made me realize that we had bigger problems than those of which we already were aware.

We knew that the big companies that manufactured cell phones, cordless phones, and other devices that could use our products were naturally reluctant to buy from small, start-up companies because they couldn't be certain that they could depend on them to fill orders in the quantities they might need, to provide the necessary quality at competitive prices—or even to be there for very long.

We also knew that manufacturers and their engineers often were set in their ways. They liked the tried and the true. The power amplifiers then in use were what we called discretes, meaning that they were made of many individual parts spread out on a circuit board. Our power amp chip would replace as many as 40 of those parts.

We felt certain that competition eventually would cause this resistance to disintegrate—if only we could get a few manufacturers to realize the simplicity, the greater quality, and the reduced size that the 2103 would provide. If a few began designing our chip into their products, others inevitably would follow.

But how do you make those initial inroads in the face of such obstacles—and even hostile opposition?

I thought I knew the answer to that, but it was not a solution that most small, new companies, hard-pressed for cash, would even consider, much less embrace, especially if their money came from VCs with tight reins on the flow.

Money was, indeed, a much bigger problem for us than we ever had imagined, astonishing proof of just how naive we had been at the beginning. We received the first $750,000 from the VCs on February 28, 1992. In little more than four months, that was pretty much gone. Although by that time we had met few, if any, of the stipulations that the VCs had placed on getting the second half of the investment, they went ahead and gave us the money in July anyway.

Before the year was out, we well understood why the VCs had been referring to our first $1.5 million as the "seed round," because we were again out of money. And it was clear that we would need even more than we already had received.

The VCs anted up another $1,750,000 in December, although by that time we all knew that we would take in only a fifth of our anticipated revenue from sales and engineering for the year. Although our portion of the company dropped, the investors paid more for the shares they received this time, thus increasing the overall value of our stock.

In the meantime, we had been growing, adding people and equipment. In June 1992, we rented the space adjoining ours at Friendly Road Business Park, module C. We knocked a hole in the wall, installed a doorway, and suddenly had twice as much room. We leased automated testing machines and began production of our early chips, for which we were getting a satisfying number of orders.

I even had an office of my own for the first time—our former conference room. The new conference room was in the other module, which remained barren of jungle-like growth. That was because Bill moved the palm tree into his office.

As 1993 began, we rented another adjoining module, B, knocked another hole in the wall and tripled our original space, making room for still more people and equipment—and, of course, requiring more money.

By spring, as we were trying to entice manufacturers to acknowledge the comeliness of the 2103, we became keenly aware that we would have to get at least twice as much cash as previously had been invested—maybe more.

Time, and not much of it, would shock us with the reality that we actually would need far more than we could have imagined earlier. Our investors had recognized this reality all along, but now they wanted reinforcement: others to share the risk.

This time, to my great relief, the responsibility of finding those new investors would not be mine. That arduous task would fall to our new CEO, Dave Norbury.

That our VCs were still with us and continued to be willing to put up more money, even if they did want help, was clear evidence that they now were as dependent on us to get their money back as we were dependent on their continuing flow of cash to keep us going and growing.

All of this gave me hope for the plan I had been plotting for some time to overcome our marketing problems: a major advertising campaign.

Well, major is a relative term in this case, but major for a company of our small size with, to this point, such modest sales for our limited and highly focused products. I wasn't thinking of TV spots that cost millions of dollars and featured Michael Jackson moonwalking with his hair on fire. After all, we weren't trying to sell Pepsis or pizzas to the masses, just microchips that would be of interest to only a relative handful of manufacturers who made the phones and other products that we knew the public soon would demand.

Nonetheless, I was thinking of spending scores of thousands of dollars, which I knew would be needed to create the impression I wanted to make within the wireless industry. And I wasn't sure how our frugal board would view that, considering our ever-tenuous financial situation, especially when they saw what I had in mind. For that reason, I wanted to be well prepared when I brought the matter to their attention.

I first had approached the board about advertising nine months earlier in the summer of 1992. All that I wanted then was approval for a modest budget that would allow me to do the essential things required for any company: a brochure and a catalog.

At that time, I had been taxing my brain for months to come up with a slogan to use in those materials that would encapsulate what our company was about. I was constantly jotting down phrases, but nothing seemed to click.

On my way to that board meeting, I stopped by my dad's house for a few minutes, and while we were talking, suddenly, without rhyme or reason, three words popped into my brain, as if they had been beamed by some benevolent radio wave: optimum technology matching.

While I am fully aware that something like this would be unlikely to happen to anybody other than a marketing director for a high-tech company, I instantly knew that this was what RF Micro Devices was about. Whatever need a customer might have, we would find the technology that best suited it and match it to that product. Three words said it all. Perfect.

As advertising slogans go, I acknowledge that this may not quite be in the league of, say, Kentucky Fried Chicken's "Finger-lickin' good," or the Yellow Pages' "Let your fingers do the walking," but for a technology company that doesn't have to concern itself with snaring ultimate consumers, or worry about what they do with their fingers, other than punching numbers on wireless phone pads, I thought it would work.

Our board agreed, and Optimum Technology Matching, which we later trademarked and still use, became the focus of our first brochure and catalog. The board granted me money enough to get them produced, and I hired an ad agency in Raleigh to design and print them. By the spring of 1993, we were sending them out to prospective customers.

These marketing materials provoked our first news coverage. Jack Scism, a business writer for the *News & Record* of Greensboro, did a feature story in the "small business" section about how we got started. It appeared on May 10 under the headline, "Silicon Valley business finds home in Greensboro." It was accompanied by a large photo of Clarence Oliver tuning testing machines, and a smaller photo of Bill, Powell, Dave, and me, looking like a rogue gospel quartet caught coming out of some bar after a prayer meeting performance. We were standing at the entrance of our first module, with only the small RF Micro Devices sign alongside to redeem us.

I was especially pleased with this article, however, because it referred to me as "master marketer Neal," a line that, for a brief moment anyway, I considered adding to my business card.

The agency that produced our catalog and brochure had prepared similar products and ads for other technology companies, which was one of the reasons we hired it. So when I began devising the big advertising campaign that I wanted to unfold in the trade magazines, I talked to the agency about what I wanted and asked its designers to come up with some ideas.

The ads they produced looked and read like every other ad in the trade magazines, as if they were written by engineers for engineers. I don't mean to imply that engineers are boring, because most I know are anything but that. Yet these ads could have given that impression.

This was not at all what I had in mind, of course. I wanted something different; something bold, audacious, humorous; something that would attract the attention of every person who opened that magazine and not only make each remember what he or she saw and read but act upon it—and quickly.

When the agency defended its approach, I knew that this was not the group for this project. And Linda Duckworth and I began a search for a new agency.

The one on which we settled, the Powers Group in Greensboro, was an even newer company than ours. It had been started just a few months earlier by Larry Cecchini, who formerly had been marketing director for a local hosiery firm, Kayser-Roth.

What impressed me most about the Powers Group was the person who actually would be creating our ads. Mike Cecil was a filmmaker who had been a partner in his own ad agency, Bouvier-Cecil in Greensboro, but he had agreed to help Larry with this project. Mike and I not only had compatible senses of humor, but his mind worked in quirky ways that I admired and enjoyed.

He immediately grasped my vision for this campaign, and we had a lot of fun kicking around ideas. The first ads that he produced were exactly what I had in mind, and I enlisted Mike's and Larry's help in springing this scheme on our board at our meeting on June 30.

This would be an aggressive campaign, I told the board. Its goals were numerous. First, we had to let people know that we existed, and secondly, we had to establish an image for our company that would set us apart from everybody else. We wanted to make other companies aware that we had a commercial technology, gallium arsenide HBT, that nobody else had, and that we intended to become the leading supplier of integrated circuits for wireless devices.

To overcome the reluctance of big corporations to deal with small companies such as ours, we needed these ads to make us appear to be much larger than we actually were, I pointed out, and by the time people began to realize that we weren't a big company, we would be one.

Moreover, I noted, the ads not only would allow us to attract new customers, but new investors as well. This really seemed to catch the board's attention.

But nothing caught their attention quite like the dramatic unveiling of the first ads we had planned. I don't think the board would have been more rapt at a Victoria's Secret fashion show—well, that might be a bit of an overstatement. Anyway, I could tell by their expressions that they liked what they saw.

Afterward, I let Larry and Mike explain the logistics of the campaign—where the ads would appear, and what the cost would be.

The board couldn't have been more enthusiastic. They immediately approved the campaign, as well as the budget for the beginning stage of advertising, which would reach $74,000.

Our first ad appeared in the August issue of *Microwaves and RF*

Magazine. It was what is known in the trade as a double-truck ad, meaning two full pages side by side. It was in full color and featured an illustration by Greensboro artist Jerry Dillingham.

It showed 80 yards of a football field, one end of which had been ripped from the earth, goal posts and all, and was rising dramatically into the air. The infield bore the RF Micro Devices logo, which I'd gotten another Greensboro artist, Vinnie Cannino, to design even before I joined the company in 1991. Enormous replicas of RF Micro chips were spread across the first 30 yards of the field, the power that was lifting it.

"Unfortunately for Our Competition," read the bold headline, "We've Just Unleveled the Playing Field."

"Leveling the playing field" was a politically correct buzz phrase at that time, and many big companies had latched onto it, despite the reality that no business wants a level playing field but one that is tilted as much in its favor as it can get. Sticking a figurative thumb in the eye of that deception was one of the things that pleased me about this ad.

In the forefront was our team, dressed in red, ready to rush onto the field and race to inevitable victory. On the other side was our competition, blue in blue, huddled in awe and fear. Two of the players were even holding hands, a little touch that was not obvious but that I hoped people might notice.

"The politically correct talk today is that 'we have to level the playing field so we can compete fairly,'" the copy read. "At RF Micro Devices, we always play fair, but now, because of our Optimum Technology Matching, the playing field, especially the RF IC playing field, will never be level again. And since it's our field, we know how to play it perfectly."

It went on to point out that we used three different technologies, including gallium arsenide HBT, to create our integrated circuits, which were "so cutting edge...that no one else even comes close.

"We know that's a pretty strong statement. But when you do something that's not politically correct in these times, you'd better have the goods to back it up."

And, of course, we did. And we invited one and all to check them out.

The issue in which the ad appeared also contained technical articles that Bill and I had written about several of our products.

This package had precisely the effect that I thought it would. The reaction was immediate. We began getting calls from companies big and small wanting catalogs, samples, more information, and they just kept coming. Old friends in the business and even marketing people at other companies called to congratulate us on the ad. (Later, when I was visiting other companies, I would even spot copies of this ad pinned on the walls of work cubicles.) Without doubt, people were suddenly taking notice of RF Micro Devices, and wondering how this company had risen so prominently from nowhere.

And our campaign was just getting started.

We sprang the next ad a month later, this time in *Wireless Design and Development Magazine*, although it also would appear a month afterward in *Microwaves and RF Magazine*, as well as in several specialty publications. It continued the sports motif.

This, too, was a double-page spread, but you had to turn the magazine sideways to read it. One full page was covered by the bottom half of a baseball, painted in excruciating detail by Jerry Dillingham. Not only could you see the texture and tiny gashes in the hide of the ball's cover, you could see the fibers of the thread that was used to sew the cover, which bore our logo, and beneath that "HBT." The ball was coming straight at you, only a split-second from your head.

"The Last Thing Our Competition Sees Just Before Being Forced Out of the Box," read the bold lines beneath the ball.

Here is the copy that followed:

"HBT stands for heterojunction bipolar transistor. But to us it stands for HardBall Technology because at RF Micro Devices that's the way we like to play the game....

"Some of our competition has been saying that HBT is experimental and not ready for market. You will note, however, that those companies don't have HBT. We have tested it, proven it, and are selling to major suppliers of wireless equipment. And it's proving itself in the field beyond everyone's belief. Though we're new in the game, we like to play hardball with the big guys. And when you play hardball, the rule is you have to play to win."

The ad was just a part of our hardball campaign. We also had hundreds of HBT baseballs made. They came in boxes illustrated with peanut shells, baseball tickets, and hot dogs, labeled, "What's at the end of a very long home run." The balls were nestled in patches of artificial grass of the sort that comes in Easter baskets.

We mailed these to key prospective customers with a letter that said, "Here for you to enjoy is your very own RF Micro Devices baseball with HBT....Look into HBT and all of our products and see how they fit your game plan. We guarantee this, when you play with our ball, you'll hit a home run every time."

The response to this obvious gimmick was phenomenal. We began getting calls from all over asking for these balls. By this time, I was regularly manning a booth at all of the trade fairs that had anything to do with our business, and I was besieged by people wanting these baseballs. We eventually had to have thousands of them made.

People even asked Bill, Powell, Dave, and me to autograph them. The demand for these balls was so great that we began wondering if we shouldn't have gone into the baseball business. We were beginning to feel like Ted Williams or Joe Dimaggio—or at least Yogi Berra. I won't be surprised at all if some day I walk into the Baseball Hall of Fame in Cooperstown, New York, and see one of these balls on display autographed by the four of us—the first baseball used in a high-tech pitch.

The hardball ad would even be translated to appear in a magazine in Japan, a country where baseball is as popular, or maybe more popular, than it is in the United States.

When we first mapped out our advertising campaign late in the spring, one of our early ideas was to stage a major press conference, perhaps in New York or Chicago, to announce that we had this advanced technology that no other company could claim. We realized, however, that a press conference about HBT and a new integrated-circuit power amplifier wasn't likely to draw much of a crowd, or get us the kind of attention we wanted. We lacked an angle, or a product, to hang it on, something with which the average person could identify.

As happenstance would have it, even as we were discussing this problem, our angle was about to present itself.

John Harwood, and his daughter, Beth, a sales engineer, had a high-tech sales rep agency that covered New Jersey and New York. They once had represented Analog Devices, until the company developed its own sales force.

Bill, Powell, and I knew the Harwoods well. We had used John as a reference when we were trying to entice money from the VCs to get the company going. He had vouched for us on several occasions. When I began to organize our sales team, the Harwoods, of course, were among the first I wanted to enlist.

One of the Harwoods' clients was a division of AT&T in Holmdel, New Jersey, which designed the company's cordless phones. Beth, who was very sharp, knew that AT&T was attempting to create the ultimate cordless phone there. This phone, for reasons unknown to us, was dubbed the Dragon Phone.

Early in the year, Beth dropped off some samples of the 2103 at the plant, hoping that it might find a place in the new phone. A few months passed with no response, but, as Beth later recalled, late in the spring, Bob Malkemes, who was in charge of getting this phone designed and built, dropped into the lab one Saturday to find his engineers having problems with the phone's discrete power amplifier, an assemblage of components that weren't quite doing what was expected or needed.

"Have you tried that new IC I sent you?" he asked, referring to our 2103.

No, they said, they hadn't, and they had to search around to find it.

"Well, plug it in and see what happens," Bob told them.

They did, and, of course, the phone not only worked but produced far greater results than they could have imagined. Bob was so impressed that he set about convincing his superiors that the 2103 should be designed into the Dragon Phone, even though it would cost a couple of bucks more than the discrete power amp that wasn't delivering.

At our board meeting on the last day of June, at which I unveiled the advertising blitz that I hoped to unleash in August, I was able to report that we had a good chance to get the 2103 designed into an AT&T phone that would bring us half a million dollars, possibly far more, in sales each year.

When Beth called not long afterward to tell us that Bob had gotten management's approval to use the 2103, we were ecstatic. There was no bigger name in electronic communications than AT&T. And what greater endorsement for our new technology could we hope for than to have AT&T include it in a new phone for which it had big plans? I quickly called all of the board members to tell them of this major development.

The original 2103 chip was packaged in ceramic, but we also encased it in plastic and called it the 2103P, and that was the version that AT&T wanted. In July, AT&T ordered 200 chips to be used in prototype phones. They requested another 2,400 in August, followed by 4,100 more in early September. On September 13, they booked a whopping order of 100,000—and two and a half weeks later asked for 4,000 more. These were to be shipped as needed by AT&T, and they gave us estimates on the numbers they would need and the dates they would need them.

We didn't have large numbers of these chips in stock. We had to get them made and packaged, and that took six weeks, or more, for each batch. But we shipped the first large number, 4,200, to the assembly plant in Singapore in September.

By November, AT&T had completed prototypes of the Dragon Phone, and they were indeed incredible. Under the right conditions, with an ad-on antenna, this phone could be used as far as six miles from its base.

Even under less-than-perfect circumstances, with no extra antenna, calls could be made and received a mile or more away. There was no static, and the voice quality was the best I'd ever heard on a cordless phone.

As soon as I saw this phone, I knew it was a product that everybody could identify with, and that I had my angle for the nationwide press conference that I'd long been thinking about.

I conferred with a public relations giant in Chicago, Burson-Marsteller, then went to the board to ask them to authorize the $50,000 that it would cost to bring it off. They asked some questions, but they had been so impressed with the results of our advertising campaign so far that they quickly authorized it.

We decided that Washington should be the site of the conference so we could emphasize the military aspect of the technology we were touting. We would hold it at the National Press Club, not far from the White House. We would have executives from RF Micro, AT&T, and TRW to speak, along with officials from the Defense Department agency that funded the original research. We would invite reporters from all the TV networks, all the big newspapers, and all the business and technology magazines. We would have a worldwide, live TV satellite feed for those who couldn't attend. And we would conduct a live demonstration of the phone's far-reaching abilities.

My first idea for the demonstration was to have somebody climb the Washington Monument about half a mile away talking on the phone to the reporters at the press conference as he went.

I made several trips to Washington to plan this event, and on one of them, I took the phone and climbed the monument myself to see if my idea might work. Alas, the reception wasn't good enough, probably because of the thick stone walls and confined interior. I was sure that if I could get through one of the observation windows and shinny up to the point on top, I'd have a clear line of sight and a perfect signal. That might be a little much to ask for the demonstrator, I figured, but it probably would have gotten us a lot more attention. Where, I asked myself, was King Kong when you really needed him?

Since the chances of finding a King Kong replacement seemed slim, I decided that we'd better restrict our demonstration.

Our press conference was held on the morning of December 7, Pearl Harbor Day, although that had nothing to do with it. We got a good turnout of reporters, mostly business, technology, and military writers from a variety of media. Bill spoke briefly about how the future of communications would depend on wireless devices, and how we had adapted a military technology that was finding its first commercial use in this new cordless phone.

Bruce Gerding, a vice president of TRW, spoke about how critical this technology was, before O.J. "Ben" Benjamin came forward to show off the phone, which was black, large for cordless phones nowadays, and a bit clunky. Ben was director of wireless terminal research and development for AT&T. He talked about the advantages of digital technology and how our 2103 had given this phone a range and quality far greater than that of any cordless phone ever made.

In answer to questions, he said that the phone would be available in January, first in AT&T stores, later in other outlets, and that the price had not yet been determined.

"I'll tell you, it will be competitively positioned," he said, going on to note that it would be more expensive than current cordless sets.

The time that I was sweating came when Ben announced the demonstration. Bob Malkemes, the engineer who'd gotten the 2103 designed into the phone, took the handset and left the room. Ben explained that Bob would proceed down a stairway to the next floor, walk to the elevators on the far side of the building, and descend to the lobby, 12 floors down. Bob was listening to all of this as Ben talked over a standard line phone.

"Bob, how are you doing?" Ben asked.

"I'm doing real good, Ben. I've crossed around the other side of the building here, going down the hall. I'm just outside of the elevator right now, and I'm going to proceed down to the lobby."

The two TV monitors in the room now showed the glass- enclosed elevator as it began its descent, Bob clearly visible with the phone to his ear.

"I'm passing the fourth floor right now, headed down," he said, as the elevator zipped past huge Christmas decorations and lights.

"Well, you're on the elevator right now?" asked Ben, who couldn't see the monitors.

"That's right."

"Wonderful. People must be wondering why you're taking your phone with you. I would imagine most people would believe it's a toy of some sort."

The reporters chuckled in response.

"I think they're pretty impressed," Bob said, as the elevator reached the shopping arcade. The door opened, and he stepped out smartly into the Christmas glow. "I'm down in the lobby right now."

"Oh, that's great," Ben said, turning back to the reporters. "One of the things I'd like to bring to your attention right here, in addition to the range, is the fact that the voice quality as you can hear is quite good. The phone is protected as far as eavesdroppers is concerned, and the digital technology over time will allow us to do things related to data that is not possible in today's cordless phones.

"So this is a significant milestone for us at AT&T and hopefully for the nation's being able to compete in this marketplace. It represents a significant move forward."

I was relieved when the press conference was over. It had gone exceptionally well, everybody agreed, and now I was left only with the agonizing wait to see if it would have any effect.

It did, and even more than we had hoped.

That evening I got several calls from home telling me that we had received our first exposure on local TV news on Greensboro station WFMY, Channel 2.

"A Greensboro company may have worldwide impact in the next few years," intoned anchor Cindy Farmer. "RF Micro Devices has signed a compact with another company that would merge both companies' state-of-the-art military technology."

The screen switched to scenes taped earlier that day in our lab as Farmer went on: "It would be used to make a new line of radio circuits like those used in cordless telephones and security systems. With this technology, officials say such products would have four times the range of the ones used today."

I wouldn't get to see that until I viewed it on tape after I got home. But I was watching in my hotel room the next morning when we landed our biggest score.

On a day when the news was dominated by the random shooting of train passengers going home from work in Chicago the previous evening, ABC's *Good Morning America* also featured a segment on *Fortune* magazine's hot new products of 1993, which ranged from the Clinton administration's health plan to a glow-in-the-dark Timex watch.

"Not on *Fortune*'s list," said reporter Stephen Aug, going on to an item that we considered to be of far more significance, and, we hoped, durability, "is this new cordless telephone unveiled yesterday."

He was holding up the Dragon Phone, although he didn't call it that.

"Unlike existing wireless phones, this one has a range from a mile to a mile and a half from your home. It uses military technology developed by RF Micro Devices and TRW. Made by AT&T, and it goes on sale next month for about $400. Just how many people will spend that much for a wireless phone? Well, maybe they ought to take a survey—or in this case, a telephone poll."

We were so overjoyed at making national TV that we didn't pay much attention to Aug's awful pun, or to the question he raised, but it soon would come to haunt us.

The Dragon Phone did not become available in AT&T stores in January, as had been promised at our press conference. We knew that production delays sometimes happened, and we weren't concerned because we still were shipping chips to the assembly plant in Singapore.

Any doubts that might have surfaced about the phone's failure to appear were erased on January 18, when AT&T booked orders for another 176,000 chips, far more than we had anticipated. That would bring us nearly $700,000 on top of the nearly $450,000 for the previous orders, and we were in a celebratory mood.

When the phone still didn't appear in February, we began to hear rumors about problems at the assembly plant, but we figured it all would work out because AT&T was still requesting chips. We shipped 15,400 that month, and continued shipping into March.

I was in San Jose in April visiting customers when I got a call from Vic Steel. Vic had joined us as a design engineer a year earlier, but later had agreed to become applications engineer for marketing. We worked closely together.

"Jerry, I've got some bad news," he said, going on to tell me that Beth Harwood had called to tell him that AT&T had cancelled the order for 176,000 chips.

Clearly this was trouble, but it would take a while for us to realize just how bad it was. We soon would learn that not only had production of the Dragon Phone been stopped, but that the phones already manufactured never would reach market. We never really found out why, other than that the phones were so expensive to make that the retail cost, at least $400, as Stephen Aug had reported on ABC, was considered to be more than consumers would be willing to pay.

This was a severe blow to us, a dark day that cast a pall over the company. We'd had such confidence in our chip and in this incredible phone, and we were certain that this was going to be our first major success. But suddenly the millions that we had anticipated receiving from the Dragon Phone had been snatched away.

To make matters worse, AT&T soon informed us that it wanted to return 45,000 of the 60,000 chips we had shipped so far, and I would have to spend many weeks haggling and compromising before that matter was resolved.

Still, we had absolute faith in the 2103, and another major, long-established company already had designed it into a series of products that should bring us far more income than the Dragon Phone ever could have produced. Beyond that, we had just snagged a deal with a much newer company that would have far-reaching consequences.

6
Syncopal Episodes

Curtis Griffin couldn't have come into the life of our company at a better time. Curtis was a young engineer, just turning 40, who worked for Motorola at a technology center in Plantation, Florida. He had a great job, and he loved it.

His specialty was power amplifiers for the two-way radios that Motorola made primarily for police and emergency services all around the world. His job was finding emerging technologies that would make these radios better, faster, and cheaper to manufacture. He loved his job because he got to visit all the major technology labs in the world, meet top scientists, and see before others what the future held in his field.

Curtis saw one of the product announcements for the 2103 that we had placed in the trade magazines and was intrigued by it. In the spring of 1993, he called me and said he'd like to come see the company and get to know more about us and our power amp. I don't think he had any idea of just how small an operation we actually were, and I had no intention of warning him, because I didn't want to scare him away.

The time he chose to come, in mid-April, was not a good time for a visitor who wanted to spend the night in the Piedmont Triad, as our area of Greensboro, Winston-Salem, and High Point is known. It was during the spring home furnishings mart, centered in High Point, which attracts buyers from all over the world, fills every hotel and motel room within a 50-mile radius, and keeps locals out of restaurants for two weeks.

Curtis came with his boss, Ken Hansen, and with a friend and fellow engineer, Joey Ooi, whose nickname reflected his family name—Double-O-One. They had a heck of a time finding rooms, but they didn't regret coming when they did, because the azaleas and dogwoods were in full bloom, and few places are more beautiful than Greensboro in April.

Ken knew one of our engineers, my friend Pete Bachert, who previously had worked for Motorola, and we had a great visit, although

Curtis and his group did seem a little surprised at our size. We later heard that some at Motorola referred to us as that "garage-door outfit" in Greensboro because of the roll-up doors that were in the back of each of our units, but that didn't matter because what we had developed behind those doors was something that Motorola, and lots of others, could use.

Actually, Curtis later said, the thing that had impressed him most about us was that we seemed so Southern, so polite and well-mannered, and that mattered to him. It mattered to us, too, and we took it as a splendid compliment. We also were impressed by Curtis. He had a great personality, a lively and funny guy who really knew his stuff. He was one of those people you hoped to land as a customer just so you could spend time with him.

At this point, Motorola's two-way radios were using a discrete module for the initial stage, or driver, of the power amplifier. It was made by a division of Motorola and cost about $20. Curtis was certain that an integrated circuit would be faster, more efficient, and less costly. That was what prompted his interest in the 2103.

Motorola was selling about a million of these emergency two-way radios worldwide every year at prices ranging from about $500 to as much as $4,000. They were rugged devices, designed to last 10 years and more, and the nature of their work demanded that they be absolutely dependable. We thought that the 2103 fit those requirements exactly, and we hoped to convince Curtis and his bosses of that.

Curtis took some sample chips back with him and discovered that their performance was far greater than he had anticipated. They could cut the current required for these modules by half. The problem for him was that the 2103 was priced higher than he wanted to pay, although we wouldn't learn that until much later.

Curtis' plan was to get Motorola to design its own integrated circuit driver for the power amp, which he thought would create a considerable savings. In June, he took his idea to the company's Semiconductor Products Sector in Phoenix, Arizona, only to be told that the chip would end up costing $5 or $6.

"Geez," Curtis said to the person then in charge of the sector, "I could buy a part from RFMD for about three." (Actually, it would be closer to $4.)

Years later, Curtis clearly remembered his response. "He looked me right in the eyeball and said, 'Go buy it.'" Curtis laughed. "I was young and stupid at the time, had more courage than insight, and I just did it."

We, of course, couldn't have been happier when Curtis called that summer to tell us that he was planning to design the 2103 into all of Motorola's two-way emergency services radios. And later, when Curtis' decision was saving Motorola $10 million a year, and making their radios more efficient and durable than ever, nobody there would dare think of him as stupid or less than insightful.

By September, Curtis had assembled a team of engineers to begin working on the design of the new power amp driver module that would include the 2103, and we anticipated beginning to ship chips to Motorola in large quantities early in 1994.

We now had two of the biggest and most established companies in electronic communications, AT&T and Motorola, as our customers, both brought to us by the 2103, the chip that had risen out of failure. We thought that we were well on our way to the big success we anticipated and that the 2103 would be the instrument that would bring it to us.

The collapse of the Dragon Phone in the spring of 1994 sent our spirits thudding back to earth, although the sting of the loss was salved by the shipments of the 2103 that by then we were beginning to send to Motorola, as well as by a contract that, unbelievably, we just had won. This contract had nothing to do with the 2103 but with a whole new cell phone technology that actually would prove to be our company's deliverance, although at many points in coming years, we would think that it might prove to be its undoing.

To understand the next part of this story, it's necessary to know that cell phone transmissions are made by several basic standards. Two were used in this country at that time: frequency division multiple access (FDMA) and time division multiple access (TDMA). A third, code division multiple access (CDMA), was about to be launched.

FDMA requires that each call be sent over a separate frequency. This method was used primarily for the early analog phones. Although it can transmit digital signals, it is a highly inefficient means of doing so.

TDMA sends signals in timed pulses. It is used for digital phones at higher frequencies than analog and is faster and more efficient than FDMA.

CDMA requires even higher frequencies, and is much faster and more commodious than TDMA. It assigns a code to each call, breaks it into parts that are sent over all available frequencies, and reassembles it at the other end. CDMA is based on spread spectrum technology that was developed by the military in World War II to thwart enemy signal interference.

Another transmission method called Global System for Mobile Communications (GSM) was developed as the standard for Europe in the early 1980s, well before cell phones became common in the United States. It is a hybrid using both FDMA and TDMA, but it employs TDMA in a different way than in this country. It is used throughout Europe, Australia, and in many other parts of the world.

These standards are constantly evolving, improving, and adopting new names as technology changes.

All require different telephones and different equipment for transmitting and receiving. A person switching from a TDMA company to a CDMA company would have to buy a new phone. But even a person with a TDMA phone still couldn't use it in areas where GSM, which uses TDMA, is the standard, because of the differences in the transmission methods.

CDMA was developed by a company called Qualcomm, which was founded in San Diego in 1985 just for that purpose. The Federal Communications Commission established frequencies for CDMA in 1993, but it would take two years for the first systems, licensed by Qualcomm, to be put into operation.

Qualcomm had high hopes for this technology and intended to profit from it not only by license fees but by manufacturing the telephones and equipment that the technology required. The company had entered a partnership with Sony to build new CDMA phones, and that's where we would come into the picture.

These new phones offered great promise for a company such as ours, which was well ahead of others in developing the highly complex integrated circuits that the phones and transmission systems required. So it probably was inevitable that we would come to have a close relationship with Qualcomm, which had established a cell phone standard for the future.

Funny how these things happen, but the beginning of this relationship went back to my less-than-successful fund-raising trip to California soon after I first came to work with Bill and Powell. Who could have predicted that that trip would provide the connection that would prove to be a breakthrough for our company three years down the road?

That link came in the person of Kevin Kelly, the sales rep for Cain Technologies with whom, by happenstance, I had breakfast on that trip, and with whom I had been so greatly impressed. Cain Technologies had agreed to represent us, and Kevin had become our rep in southern California, where Qualcomm was one of his clients.

From the beginning, Kevin was determined to hook us up with Qualcomm. In the fall of 1992, he arranged for Bill and me to fly to San Diego to attend a seminar for potential suppliers where we learned about CDMA and how it worked.

We were quick to see the potential that it offered, and we returned home with new determination to make Qualcomm one of our major customers. We began sampling standard parts to the company shortly afterward, and Kevin kept in close contact, looking for possibilities that we could seize.

Nearly a year later, in September 1993, Kevin informed us that

Qualcomm was about to open bids on parts for two different cell phones it was planning to produce with Sony.

One was a standard CDMA phone. The other was a next-generation cell phone that would provide much more than just two-way conversation.

This second phone was classified as a PCS (for personal communications services) phone. It would operate at high-frequency and allow for many services not available on standard cell phones, such as call waiting, caller identification, and call answering. Eventually, PCS phones would allow instant messaging, computer connections, e-mail, digital cameras, and many other miraculous possibilities. The FCC was expected to allocate a separate spectrum of the radio band for PCS phones in 1994, which it would in fact do, opening the way for all the wonders we see in cell phones today—and for us the potential for incredible business.

Qualcomm wanted bid proposals for the CDMA phone in October, and planned to open bids for the PCS phone sometime in 1994. We were invited to offer a proposal on a set of five chips for the CDMA phone. These were transmitter and receiver parts: a low noise amplifier/mixer, automatic gain control amplifier for both receiver and transmitter, an upconverter, and a power amp.

We submitted our first proposal before the October deadline, offering to use our full spectrum of Optimum Technology Matching: gallium arsenide HBT, gallium arsenide MESFET, and bipolar silicon.

Early in December, we were notified that we were in the running. We had a chance, and we thought we might get one or two of those chips, certainly no more, and we began working in earnest to that end.

Early in January, Bill redid our proposal, offering to design all the chips in gallium arsenide HBT, a technology that few other companies could claim. Kevin had warned us months earlier that Qualcomm was skeptical about HBT and that it would take a lot of convincing to get them to accept it, but we thought that we could do it, based on the performance and reliability of the 2103.

We had several reasons for switching to HBT. It would make design easier and give us a better chance of meeting the stringent specifications that Qualcomm required. It also would save us money in the development stage. Producing wafers for prototype chips is expensive, and this would make it possible for us to put all the chips on a single wafer.

Qualcomm had a reputation as a tough company to please, and some prospective vendors didn't bother to bid for Qualcomm work because of it. We were hungry for business, though, and willing to tolerate whatever difficulties might ensue.

We were not long in receiving a taste of just what we were getting into. I was on the phone several times a day with Qualcomm officials, answering questions, providing information. We supplied them paperwork by the ton.

Near the end of January, I arranged for a team from Qualcomm to visit the foundry at TRW-Space Park in Redondo Beach. Ten days later, I received a letter from Qualcomm's director of quality and reliability control, Joe Hypnarowski, containing a long list of "concerns." Their primary worry was whether TRW could supply "quality and reliable product, to agreed upon schedules, on a continuous basis."

The Qualcomm team thought we had not been keeping a close enough eye on TRW's processes and wanted us to establish a foundry monitoring program. Among many other things, they also wanted reliability data on HBT.

We set about the ardent task of satisfying all of Qualcomm's requests, and three weeks later, faxed Qualcomm answers and our plans for a foundry monitoring program. But that wasn't good enough.

On March 3, I received a sharp letter from Hypnarowski saying that our efforts demonstrated that we were "still lacking a firm commitment to adequate Foundry Control....Progress is less than anticipated....If you do not have dates in place NOW, you will not be ready for production...."

He wanted firm dates and quick answers. We figured that we had little chance of getting Qualcomm's business at this point, but we once again set about attempting to please the company.

Imagine our surprise, when a week later, I received a letter of intent from a senior vice president of Qualcomm announcing that we had been chosen to supply four of the five parts on which we had made bids for the new CDMA phone. The only part we didn't get was the power amp, our HBT specialty.

This was a major breakthrough for us, if only we could meet all of Qualcomm's demands.

Under this agreement, we would be paid $160,000 in engineering fees and begin design immediately. We had to have prototypes ready by July 1. The chips would have to be tweaked and working to specifications by September 1, and production quantities would have to be in Qualcomm's possession no later than November 16.

Qualcomm forecasted a need of 75,000 of each chip in 1995 and 240,000 of each in 1996. The deal was conditioned on our providing a satisfactory resolution to all of Joe Hypnarowski's concerns by April 30. The letter of intent had arrived with a lengthy copy of Qualcomm's own "Quality Assurance Operations Procedures."

We thought it unlikely that we, or any other company, could meet Qualcomm's rigid specifications under this schedule. But we signed the letter anyway—and redoubled our efforts to please Joe Hypnarowski by the appointed date.

• • •

The pending deal with Qualcomm was especially pleasing to our investors, because we were far from meeting our own forecasts for production and sales. Much to our chagrin, this had become standard operational procedure for us.

We were so far off our predictions in 1993 that, at our board meeting in September of that year, I attempted to leaven the situation with a bit of humor.

I presented our VCs with a medical release form absolving me and all company associates from any symptoms of hyperventilation, hypertension, hypotension, syncopal episodes, prostate discomfort, nausea, diarrhea, or constipation that they might experience on hearing our marketing reports. They chuckled appreciatively, but I noticed that none signed and returned the forms.

This was no joking matter with any of us, of course. Our expenses and cash outflow kept expanding dramatically. In business terms, this is known as cash burn. It was always singeing us and frequently seemed about to envelop us.

Once again we needed more money more quickly than we anticipated. We had to get a bridge loan in mid-1993, and at the beginning of December, we received the new round of financing that Dave had been working on so intensely for the past year.

The amount we originally thought we'd need—$3 million—had grown to $5.75 million. Half of that came from our faithful group of VCs and associates that they allowed to join them. The other half came from three new venture capital groups: North Carolina Enterprise Fund, Carolinas Capital, and Norwest Venture Capital.

This brought us a new board member, Bob Fleming, representing Norwest, who promptly announced his full intention to be a hard case to please and handle, although he put it in more succinct and blunt terms. Total investment in the company was now $9 million, but it was apparent that more financing would be required as we continued to grow, and Dave was in the position of finishing one round only to start another.

New investment also brought the issuance of more stock, and the corresponding dilution of the existing pool. The 36 percent share of the company that Bill, Powell, and I once owned now had fallen to 12.1 percent—and it would continue to drop.

At our first board meeting in 1992, we had converted to a fiscal year that began on April 1. Fiscal year 1994 began on that date in 1993 and ended on March 31, 1994. For that year, we had taken in only $900,000 from sales. But with all that was happening, we were confidently predicting far greater revenue for fiscal year 1995, $6.16 million to be exact, with a profitable fourth quarter, which would be a first for us. Still we anticipated losses of $1.55 million for the year.

By May, however, our confidence had sagged. Although we had picked up business not included in our predictions, we were forced to lower our revenue expectations for fiscal '95 to $5.6 million, in large part due to the stunning loss of the Dragon Phone.

Every chip we were making had to be sent away for packaging, the process that attaches the connectors that allow the chip to be plugged into a circuit board. The packagers all were in Asia. They returned the chips to us for testing before we put them on plastic reels and shipped them to customers.

The purpose of testing is to make certain that every chip is performing to expectations. The manufacturing process causes variances in chips. Chips made side-by-side in the same wafer may not provide the same results. One wafer may produce far more working chips than another. Those chips that don't meet specifications have to be identified and eliminated.

Our testing facilities were considerably cramped and overburdened at this time. The surge of business that we expected from Motorola and the forthcoming Qualcomm deal added urgency to our need to expand our testing operations with people and new, leased equipment.

On April 1, we had rented two more adjoining modules at Friendly Business Park. These were in another building a short distance from our present three modules. Through the spring and early summer, Powell, Bob Hicks, and Fred Adkins, our quality control engineer, were working hard setting up our new testing facility in those two units.

By mid-summer, somewhat behind schedule, we completed the design of the four Qualcomm chips, and in late August, we sent the prototypes off to TRW to be fabricated. We got them back on October 3, and to our immense relief, they all worked. As usual, they didn't work to Qualcomm's specifications, but they worked. It only would be a matter of fixing them. Just how long that might take, we couldn't be certain.

By this point, we were quite a different company than we had been a year before. We now had 45 standard products, as opposed to only 16 a year earlier. We had 36 employees, compared to 26 at this point in 1993.

A year earlier, we had no quality monitoring program for foundries. Now, thanks to Qualcomm, we had one. In October 1993, our testing lab was small and cramped. Now it was greatly expanded with new equipment and more people.

A year earlier, we had been counting heavily on AT&T's Dragon Phone and on Motorola's two-way emergency radios as the forthcoming mainstays of our business. But the Dragon Phone had disappeared. And to our great dismay, we recently had been told by Curtis Griffin that Motorola higher-ups had decided to design the 2103 out of their radios and create their own integrated circuits for their power amps and drivers.

The Motorola decision, coming only six months after we had started shipping to them in large numbers, was yet another unexpected blow. But it would take about two years for Motorola to get its ICs designed, approved, working, and fully employed. Meanwhile, we still would be shipping lots of chips to them.

By this point, we had established relationships with all the top manufacturers of cell phones, primarily because of our power amps. After the 2103 had proved itself, we had decided to make power amps a primary focus of our business. We had designed several others since the 2103, and recently had completed a far more complex one, the 2115, although we were having packaging problems with it. But there was no question that we were making the best integrated circuit power amps on the market, and that was what was attracting the top cell phone makers to us.

Motorola, which was number one in the business, was showing little interest, because it was creating its own discrete power amps and was reluctant to abandon them. But the number two company, Ericsson, which was headquartered in Sweden but had a design center in the Research Triangle of Raleigh, Durham, Chapel Hill—which Bill, Powell, and I had visited in the first year of our company, hoping to attract some business—was now interested not only in our power amps but our oscillators, as well.

The number three company, Nokia, headquartered in Finland, which within a few years would come to dominate the sales of cell phone handsets throughout the world, also was showing interest in our power amps.

On top of that, in July, Bill had submitted a bid proposal on five parts for Qualcomm-Sony's new PCS phone, and there were signs that we might get that contract, too.

The future was beginning to look bright indeed, and it began looking even better for me on Halloween Day because I had scheduled a very special job interview.

The person I was to meet that day was Kathy Adams, who currently is—and I hope forever will be—my administrative assistant. Once, Linda Duckworth, our first employee, had very ably and bravely filled that job and several others. But she had started a family, and her second pregnancy had proven to be especially difficult. She was out of work for lengthy periods, and when she finally returned, it would be part-time and in a different position.

Meanwhile, a couple of good people had substituted for her—Merci Danielson comes to mind—but recently I had suffered through several assistants from temporary employment agencies who were of little help at all, and I was nearing a state of despair.

I was to some degree helpless without a strong and dedicated assistant, because I may well be the last person from the old school of business who still dictates everything. I had to have somebody close-by who was adept at shorthand.

I also was somewhat limited when it came to filing. My system was to stack things up until they fell over, then pile more on top of that. With the blizzard of paper work going back and forth between us and Qualcomm, not to mention all of the other documents from our growing list of customers, my office had become a disaster zone with papers heaped everywhere. People would come to the door, peek in, and think I wasn't there because I was hidden behind the paper peaks.

My wife, Linda, regularly had to listen to all my complaints about my inability to find a good person to work with me. It was she who suggested that I attempt to lure Kathy from her present employer. Linda once had worked with Kathy at Alma Desk Company, where my father worked, and knew how good she was. Kathy had been an executive assistant at two other furniture factories as well, but now was training in marketing at a steel company in Randleman in Randolph County, where she lived. If I put on a real charm offensive, Linda told me, I might be able to entice her to join us.

If I'd had any idea at the time just how incredible Kathy actually was, I'd have put on the sales campaign of my life that Halloween Day.

I needed somebody who could handle almost anything; somebody dedicated, loyal, trustworthy, who could maintain strict confidentiality; somebody who was especially well organized. Kathy proved to be all of that and more.

I truly can say that I am the envy of every executive who has had contact with Kathy. Over and over, wherever I go, people tell me how thorough, precise, organized, and pleasant Kathy is, what a joy she is to work with.

Her disposition is always the same, always cheerful. And I've never worked with anybody who works harder, or is more dedicated. No matter how long I work, she won't leave until I do, unless I force her to go. And she remains on standby, with a basic set of records at hand, 24 hours a day, seven days a week. I can call her cell phone from anywhere in the world, at anytime of the day, and she will answer within a few rings. If I need the name of the president, or marketing director, of some company in Taiwan or Korea, she usually will supply it immediately, or call me back shortly with the information.

There is no situation that I can't trust Kathy to handle, no matter how stressful. Even if someone is angry and abusive with her, she remains perfectly calm and pleasant.

There are times when my life becomes so hectic and frantic that it would be unbearable without Kathy as buffer and organizer. Without her, I'm sure, I'd have been in meltdown long ago. With her, I am able to accomplish so much more.

After meeting her, I knew that I'd found the person I'd been seeking for a long time, but I was getting ready to leave for California the following day to deal with Qualcomm problems and other matters, a trip that extended over weeks. When I got back and called Kathy, it was almost Thanksgiving and a lot was going on, so we agreed to talk after the holiday.

When we did talk about her coming to work, she said that she'd have to give a notice and wanted to take a little time off before the holidays, and we agreed that she would join us after Christmas.

Meanwhile, my office was piling up to such extent that I was beginning to tiptoe around out of fear of setting off potentially disastrous paperslides. I was a little concerned that when Kathy finally came to work and got a fresh glimpse of it, she might just turn and flee.

December, however, proved to be a great month for me.

We had continued our advertising on a lesser scale during the year, with only a couple of full-page ads in trade magazines. But we closed out the year with a grand finale.

This ad would be the last that Mike Cecil would create for us before he moved on to bigger success promoting another local product, Krispy Kreme doughnuts. It appeared in the December issue of *Microwaves and RF* magazine. It featured a famous photograph, a recreation of Alexander Graham Bell's first telephone call. Bell was seated at his desk. A group of men wearing the dark, stiff business suits of the late nineteenth century stood behind him, witnesses to the great event.

But in our version, Bell wasn't talking into the original telephone. Instead, he was holding a prototype of the new Qualcomm-Sony CDMA phone for which we'd made four essential chips. The RF Micro logo was attached to the side of his desk on a gold plaque. Spread out on his desk were papers bearing the RFMD and Qualcomm letterheads.

We had removed the heads of all the witnesses and superimposed those of others. I was looking over Bell's right shoulder. Bill was on my left, Powell on my right. With his white beard and mustache, Powell actually looked as if he belonged in the original photo. We included several of our engineers—Kellie Chong, Fred Adkins, Bob Hicks, Leonard Reynolds, and Billy Pratt—along with Doug Dunn from Qualcomm, and Aaron Oki and Bob Van Buskirk from TRW.

The ad was such a hit that *Microwaves and RF* magazine chose the doctored photo for its cover. The main article featured on the cover was about our HBT chipset for Qualcomm, which the magazine's editors had chosen as one of the year's top twelve new wireless products despite the fact that the chips still weren't working to Qualcomm's specifications, a situation that was beginning to cause a great deal of stress between us.

All of this attention brought us lots of response and new business. We also were getting close on some big deals that we had been working on for a long time. One was with Ericsson, providing power amps for its phones, and I felt certain that we soon would land Nokia as a customer as well.

But best of all, to cap out the month, on the day after Christmas, Kathy Adams came to work with me. She immediately began tackling the mountains of paper in my office and restoring order and sanity to my work life. What better way to welcome a new year?

7
Go On and Shoot Up In Here Among Us

The mid-1990s were a gold rush era for the cell phone business. Only a relatively few people had cell phones, and most who were aware of them wanted one.

The business was growing by such incredible leaps, 40 percent to 50 percent a year, often more, that it seemed without bounds.

Service providers couldn't set up systems fast enough, and the demand for phones, base stations, and other equipment was overwhelming.

As 1995 began, we were so caught up in this madness that we really had little time to think about the possibilities. We were like a guy on a roller coaster being shot to thrilling highs before plunging to sickening lows, hanging on for dear life. We had an occasional glimpse of what lay ahead, but the turbulence of the moment made it hard to concentrate.

There was no getting off, and we wouldn't have considered it if we could. I've always loved roller coasters, the steeper and faster the better, and I, and all the rest of us, were committed for the ride, taking it as it came.

We were certain that 1995 would be the turning point for RF Micro, and it was an exciting time because of that. But it also would prove to be intensely stressful.

At our first board meeting, on February 15, I was able to report that business prospects looked better than ever. Orders were strong. And we were on the verge of a really big deal that Vic Steel, our marketing engineer, Greg Thompson, who had joined us in late 1993; and I had been working on for a long time—a partnership with Nokia to produce power amps for a whole series of new phones in multiple transmission methods. Vic and I would be going to Finland on March 1 to close that agreement.

It even looked as if we might get some business we previously had thought to be lost. Itron, a company that was attempting to make wireless meter readers for electric companies, was planning a $3 million order.

By our next board meeting, on March 28, three days before the close of fiscal 1995, prospects looked even better.

AT&T had just asked us to design parts for a new CDMA phone it was planning. As events would turn out, this phone, like the Dragon Phone, never would reach market. Unfortunately, this would become a pattern with many AT&T phones during this period, and eventually Motorola would take over AT&T's phone-building division.

A week earlier, a company called Cincinnati Microwave, maker of top-quality radar detectors, had chosen one of our silicon chips for a remarkable, high-quality cordless phone with similarities to the Dragon Phone. The company already had ordered 22,500 parts and anticipated needing 300,000 in the coming year. (That phone was so good that I advised my wife to invest some of her retirement savings in the company. Alas, problems with this phone would prove to be Cincinnati Microwave's undoing, eventually leading the company into bankruptcy at the cost of my wife's savings. I still haven't heard the end of it.)

We had booked so many orders in the previous five weeks that our fiscal '95 bookings ended up $1 million ahead of expectations. Actual revenue for the year, however, was only $1.6 million, well short of the $3.7 million we had forecast. But it was clear from recent developments that we finally were closing in on forecast shortfalls and were about to see their end.

One reason for this was that after nearly three and a half years of work, we finally had begun shipping in quantity to one of our earliest customers, Digital Security Controls. Getting to this point with these two silicon chips had been an exercise in immense frustration and aggravation, with frayed nerves on both sides. Sometimes our shortcomings had been the source of the problems and delays. At other times, the blame lay in DSC's discovery that the chips needed functions that it hadn't contemplated, requiring extensive redesign. But now all the problems were behind us, and DSC was quickly becoming one of our top customers.

Our biggest new development by far had to be cloaked in secrecy. That was our new partnership with Nokia. We had reached agreement at the beginning of the month, although the documents were yet to be signed.

Nokia required strict confidentiality about the relationship, as well as about all of the company's parts and orders, because its management didn't want the competition to get even a whiff of their plans.

To maintain secrecy, we had given Nokia a code name: Atlas. We called Nokia Atlas in-house for so long that even today some in our company still occasionally refer to it that way.

Under this partnership, we guaranteed Nokia our HBT technology in the volumes that it would require and granted it participation in any expansion of facilities that production might demand. We also promised to keep Nokia informed of all new developments in radio frequency technology and to give it first opportunity on new products we produced.

We had no doubt that this partnership was the most important event in our company's history to date, bigger even than landing the Qualcomm projects, although we had no idea at the time just how dramatically it actually would affect us. But we felt certain that with Nokia and Qualcomm as anchor customers, we were well situated to become the major company we had dreamed of creating.

We had been working on custom designs for Nokia chips well before we officially entered this partnership. Already we had five at TRW for prototype chips, with three others nearly ready to go. Preliminary work was underway for several more. Also the company had designed our new 2115 power amp into two phones, and we had a $636,000 backlog of Nokia orders so far.

The major question now in our minds—and it was becoming more urgent by the day—was whether TRW would have the capacity to allow us to keep our commitments to Nokia, Qualcomm, and our growing list of other customers.

Already, our amazing power amp, the 2103, was by far our best-selling standard part, and it was made at TRW. All of the Qualcomm and Nokia chips would be made there, and we had no reliable way of knowing what the demand for those might be, although it was becoming apparent that Qualcomm was going to need more chips much more quickly than either of our companies had anticipated.

Our orders at this point showed us that 74 percent of our business would have to come from TRW's foundry, which previously had run only relatively small batches of chips primarily for government use and never before had been presented with such demands.

This meant that among all of the vital issues that we now were facing, the most pressing was to make certain that TRW was willing to do whatever was necessary to meet the production that we soon would require. Yet we were well aware that TRW was an old and conservative company, with a dependency on government contracts. We knew, too, that it had a big bureaucracy that, like government, often was slow to react.

In April, I was stunned by an unexpected call from Curtis Griffin, the engineer who had designed the 2103 into Motorola's emergency radios. Some of Motorola's radios had been failing in final tests, and it suspected that the 2103 might be at fault.

We thought that was not possible. It was a proven part, completely reliable. Motorola had shipped more than 200,000 radios with the 2103 by this time, all without problems. Why would the chip be failing now? It had to be something else, we thought.

Curtis suspected a wire-bond problem, and he soon confirmed it. He sent us 10 or 15 chips as evidence, and with sinking hearts, we had to admit that he was right.

A wire bond is the final addition to a chip, and it was not a step that we, or TRW, performed. The incredibly fine wires that connect the circuits in the many layers of a gallium arsenide HBT chip such as the 2103 are of pure gold. For the chip to function, a minuscule layer of gold, called a bond pad, is spread across the top of the chip so that its circuits can be connected to other elements in whatever device it is to be used.

When the chip is packaged in plastic, ceramic, or metal so that connections can be added to allow it to be plugged into a circuit board, a machine with a snake-like fang filled with a pure gold wire only one-thousandth of an inch thick uses high heat to attach to the bond pad whatever number of tiny wires the chip requires to function. With the 2103, it was 14. All of this is done automatically by a machine that has to be set with ultimate precision.

If these wires aren't properly welded, they can pull free from the bond pad under stress and cause the chip to fail. This clearly was what was happening with the Motorola chips—and, as we soon were to discover, not just with Motorola's. A small company in Maryland that made tiny transmitters for tracking migratory birds and fish also got some of the bad chips and subsequently lost touch with some of the wildlife.

Curtis was finding a failure rate of 2 percent, or less, but that was enough to create a crisis for him. Emergency radios require a zero failure rate. Motorola couldn't afford to ship a single radio with a chip that might fail.

Yet, the company couldn't wait until we could solve the problem and get more trouble-free chips to it either. Production lines in Iowa, Ireland, and Florida were assembling these radios, and it would cost millions to shut them down.

This business was bringing more than a billion dollars a year to Motorola, and Curtis quickly found himself the object of attention from the company's top executives.

"If you screw that up, people call you at home—and they did," he remembered.

For us, there was no immediate solution to the problem. The wire bonds came loose only under intense heat, and we didn't have the equipment to test the chips to find the few that might fail. All we could do was identify the batch, or batches, of chips that contained the bad ones and find out why the wire bonds were defective. Later, we would learn that other semiconductor manufacturers had suffered similar problems with the company we had employed to package the 2103, and we would drop the company.

We could run more chips and have them packaged by another company, but it would take six to eight weeks to get them to Motorola.

Curtis had to have a solution NOW.

"There was no turning back," he said. "We had to fix the problem."

Motorola had contracted with a company called CMAC in West Palm Beach to build the power amp modules in which the 2103 was the driver. Finding a way out of this distressing fix was as important to CMAC as it was to us and to Motorola, and Curtis called on them for help.

Within a matter of days, Dan Leiwe and Randy Norman at CMAC developed a method of heat-testing the 2103 that revealed and discarded every chip with a bad wire bond. Curtis later described their efforts as amazing. This was a painful and costly strain on CMAC, but the remarkable resourcefulness of Dan and Randy allowed Motorola to continue production.

This time-consuming extra step went on until we could turn out reliable chips and get them to CMAC. We eventually had to take back more than 40,000 chips from the bad batches, 98 percent of which were perfectly sound. All, however, ended up in the dumpster. The loss to us came to $184,000, more than 10 percent of our total sales for the previous year, a difficult blow at a time when we still were spending money far faster than we were bringing it in.

Ironically, our dedication to accountability and strict honesty ended up costing us even more. We sent a person from our financial office to check Motorola's books when we were crediting the returns to make sure that the numbers were right, and she discovered that Motorola had made a $20,000 bookkeeping error in our favor. We credited that, too, against future shipments. Some people might look at that and conclude that honesty doesn't pay, but I'll always believe otherwise.

More than a year now had passed since we had won the contract from Qualcomm for the CDMA phone, and the company since had chosen us to do the chipset for the new PCS phone as well.

Although we originally were supposed to have the CDMA chips ready for mass production by the previous November, all of us had understood that was unrealistic. It normally takes 18 months to get almost any chip working adequately to meet production standards.

Tension had built steadily as time passed and the chips still weren't meeting Qualcomm's stringent specifications, and now Qualcomm had begun applying intense pressure on us.

In fairness, it should be noted that Qualcomm itself was under great pressure at the time. Although the company's CDMA technology had been chosen for cellular service by several major companies in the United States, including Bell Atlantic, U.S. West, Alltel, Ameritech, Airtouch, and Nyex, as well as companies in South Korea and Hong Kong, no system was yet functioning. Qualcomm was under heavy criticism for delays, and some were questioning whether CDMA had been adequately tested and even if it would work on a commercial scale.

It was understandable that Qualcomm would be passing some of this heat on to us, but we had begun to feel that at times the company was expecting the impossible.

We were putting in long hours and working as hard as we could to get the chips ready, but Qualcomm kept moving the goal posts on us. Its engineers would discover something else that they wanted the phone to do and change the specifications to fit that need. They constantly were trying to get the chips to do more and to impose tighter standards for us to meet.

They weren't alone in wanting changes in the specifications. We desperately desired them, because their specifications were so tight—far more rigorous than normal functioning required—that in some cases meeting them was not possible.

The specifications were an unremitting source of friction between us. Our engineers and theirs were regularly holding seemingly endless meetings about them, both in person and in lengthy conference calls. They wanted more. We needed less. And at times these conferences would become quite contentious.

We pointed out that no company could meet the specifications they were demanding, but they held firmly to their position.

This friction eventually became a process, grinding on both sides day after day, wearing everybody down. As the pressure to move to production increased, our sessions became more and more heated, the result often being little more than anger and frustration on both sides.

When we weren't meeting, I became the conduit between our engineers and managers and theirs. Almost every day, I would be on the phone for three or four hours with somebody from Qualcomm, just trying to clarify their questions, concerns, and often vitriolic complaints. Hour after hour, I'd get beaten on, and then I'd have to come back and attempt to be calm and pleasant.

"Okay," I'd say, "Here are your answers, and this is what we're going to do."

To make matters worse, Sony had sent a large team from Japan to set up assembly lines for the CDMA phone. Japan had been a big part of my territory when I worked for Analog Devices, and I had done a lot of business there, visiting regularly for years. I had come to love the country and knew and admired a lot of people in the technology business there.

Although I didn't know anybody on the Sony team at Qualcomm, I expected to get along well with them, and did to a point. But there seemed to be a lack of communication between the Sony team and the Qualcomm folks, and that created confusion.

One day I'd get instructions from Sony people that would be contradicted the next day by somebody from Qualcomm. One day I'd get berated in English, the next day in Japanese.

I sympathized with the frustrations of both groups, and heaven knows that nobody wanted to please them any more than I did, but at times it was about all I could do to dig my way out from under the bafflement and abuse that they kept heaping on me.

These days ranked among the most draining of all my work experiences, and although I probably wouldn't have believed it at the time, they were to grow much worse.

Getting Qualcomm's chips to meet specifications enough to begin production was only part of our problem. We still had to make certain that TRW could meet our needs when we finally reached that point. Late in April, we had begun a series of meetings with TRW executives to come up with a plan. TRW agreed that something had to be done, but what, when, at what cost, and with whom were the questions.

On May 10, Bill, Dave, Dean, who was now our vice president for finance, and I flew to California to try to get some answers before our board meeting scheduled for the 18th.

The Space and Electronics Group at San Diego, which had more than 5,000 employees, was one of TRW's three major groups, under which were many divisions. It was headed by Tim Hanneman. We dealt with the Electronics Systems and Technology Division, of which Dave Vandervoet was in charge. But most of our communications were with Bob Van Buskirk, who was director of gallium arsenide applications.

At this meeting, we listed all of TRW's options, which ranged from doing nothing to building a new fabrication facility—or fab, as it is called for brevity's sake—and filling it with more productive equipment than TRW currently was using. The latter was our preferred solution.

We also established "boundary conditions." These were stipulations which had to be included in any agreement to build a new fab.

TRW made it clear that it would be unlikely to provide full funding for the fab. It wanted the right to take on partners beyond us. It insisted that the gallium arsenide HBT technology remain under its control and that its investment in its development had to be a factor. TRW also wanted to be more than just a wafer supplier. Its right to move into commercial products, systems, and services of its own was reserved as well.

All of us had concluded that we needed to make an investment in the new fab in order to have a say in its operation and future. This either could be in cash, or in stock, although we made it clear that our VCs would approve no arrangement that might lead to them losing control of the company.

Whatever we worked out, we wanted to be sure that we and TRW controlled the fab's business. We also insisted that our development of HBT products and customers had to be a factor. And we wanted a guarantee that TRW would provide all future technology upgrades to the fab and agree not to set up any other foundries in competition.

At this meeting, we even established a schedule for resolving this matter. By June 1, the Space and Electronics Group would narrow its alternatives to two or three. By June 21, it would choose a course and complete a detailed business plan. TRW's corporate management would finish its review of the proposal by July 15, and final approval would come no later than August 1.

This schedule, like much else that was decided at this meeting, would prove to be utterly unrealistic.

As we struggled with these complex, major issues, another concern that had been lurking in the background was rapidly becoming more urgent. This was our increasingly crowded quarters, which were becoming a bigger detriment by the day.

Nearly a year earlier we had begun discussions about acquiring a new facility to consolidate our operations under a single roof. We had been steadily moving toward that goal, but not fast enough to satisfy our growth.

Having our operations spread through five different modules on Friendly Road had created communications problems. And we all believed that it was hurting business.

Potential customers would come to visit, see our multiple bays, our garage doors, our crowded conditions, and think that we were just some rinky-dink operation. We felt that they might get the impression that we weren't capable of handling a big job, that we had no place for future growth to accommodate their requirements. We were convinced as well that the situation also was beginning to hurt our ability to recruit top-flight professionals.

We had communicated all of this to our board, and they were understanding. In February, we laid out our options and the likely costs.

The option that would be least expensive was to stay in place and rent more modules, but that was intolerable to us. Another choice was to find an existing building for lease and move into it. We already had looked at several without finding anything with immediate appeal.

The third option was the one that Bill, Powell, Dave, Dean, and I favored. That was to pick a site, and find a developer who would buy the land, construct a building to our specifications, and lease the property to us. This would cost more than the other options and require committing to a lease of 10 years, but it offered the overwhelming benefit of allowing us to design a structure that fit our exact needs.

The board gave us permission to proceed on this course and prepare a detailed plan for a later meeting. We thought that we needed a building of 25,000 square feet, about three times our current space, and by May, we had received proposals from three developers and picked our preferred site.

The site was a few miles away on Thorndike Road in Deep River Business Park, along I-40, only minutes from the Triad International Airport, and just a short distance across the highway from the Computer Labs Division of Analog Devices, where Bill, Powell, and I once had worked.

Seventy acres were available at this site, and I was so confident that we were going to become a big company that I thought we should buy all of it so that we eventually would have room for a campus. I tried to convince the board of this, but they thought that we should stick to 5.5 acres, which would allow room for a 25,000-square-foot addition if we should need it.

At our regular meeting on May 18, the board passed a resolution committing to building on the site, and instructing us to begin deciding among the three developers, planning the building, and settling on a construction company.

Most of the strain and conflict we were having with Qualcomm had not yet reached top management there, and we were optimistic that despite the constraints being put on us, we still could work out the problems without them reaching that level. At Qualcomm's request, the company's president, Harvey White, came to speak at our board meeting on May 18. It was a congenial occasion. His presentation was about the reception CDMA was getting from service providers and the great opportunities it was offering for all of us.

His real purpose, however, was to alert us that Qualcomm was going to need chipsets in far greater numbers than anybody had anticipated, and he wanted to satisfy himself that we could provide them. Another goal of his mission was to motivate us to ratchet up the tempo to get the chips ready for production.

This squeeze from the top was only a foreshadowing, however, of the pressure that was about to descend upon us.

In 1994, the FCC had auctioned off the rights to the bands in the radio spectrum that would allow personal communications services (PCS) phones, and numerous companies had joined into consortia to bid on them. One consortium was PrimeCo, which was made up of Nynex, Bell Atlantic, US West, and Airtouch.

On June 6, just three weeks after Harvey White's appearance before our board, PrimeCo announced that it had chosen CDMA as its operating technology. The announcement brought a $5.25 surge in Qualcomm's stock in a single day. PrimeCo was licensed for 57 million subscribers, and the meaning of this to us was truly transforming.

To this point, almost all of our efforts for Qualcomm had been aimed at getting the chips ready for the cellular CDMA phones, which would have a much smaller market than the multi-use PCS phones. But we also had the contract for the PCS chips, and we were Qualcomm's only supplier for them. PrimeCo's announcement meant that Qualcomm soon would be needing millions and millions of these chips. We figured they would bring us $20 million or more annually.

As happy as we were about these prospects, this news also was occasion for chilling thoughts. We still were in the early stages of development of the PCS chips. The sometimes overwhelming pressure we already were under to get the CDMA chips in production might come to seem nothing compared to what the PCS chips would bring.

We weren't long in discovering the reality of these fears. Qualcomm immediately cranked up the priority on the PCS phone and the burden on us rose correspondingly.

Then, just six weeks after PrimeCo's announcement, we and the entire telecommunications industry were surprised by an even bigger development.

Sprint, which had purchased licenses for 156 million PCS users, also chose CDMA, despite widespread predictions to the contrary. Qualcomm's stock leapt another $6 on the day of the announcement. We only could imagine what this meant for us.

Dave put it succinctly in a memo to the entire staff: "The implications for RFMD are ASTRONOMICAL!"

Qualcomm and Nokia agreed with us that the construction of a new fab was the best way to assure adequate production of the chips that both soon would be needing, and both companies, independently, were making their feelings known to TRW.

After his appearance at our board meeting in May, Harvey White, president of Qualcomm, made several calls urging this course to Tim Hanneman, who headed TRW's Space and Electronics group. White also proposed that Qualcomm would be willing to invest in the fab.

On July 3, I flew to Dallas for a meeting that included Werner Gruber, a vice president of Nokia Mobile Phones, and TRW's Dave Vandervoet, Bob Van Buskirk, and Ed Cypert, vice president of operations. Werner made it clear that Nokia's business was contingent on the construction of a new, high-capacity fab, and the TRW people agreed that this was their preferred option.

TRW's present equipment produced three-inch wafers and couldn't be altered to make bigger ones. Equipment was available to turn out four-inch, and even six-inch wafers, which would provide far more chips at a much lower cost. That a new fab should have this capability went unquestioned.

Tim Hanneman seemed to be enthusiastic about building such a fab, and soon after our meeting in Dallas, he began forwarding reports to corporate officers urging them to "get this thing moving faster."

In the meantime, he informed us that TRW was willing to make an $11 million expansion of the current foundry, which included buying new equipment and hiring 50 more people.

Dave reported all of this to our board on July 10, offering his opinion that the probability was high that TRW would build the new fab, although the timing of corporate approval was unpredictable.

"We need to be patient," he counseled.

Dave also warned the board that we should quickly come up with new financing sources if the board agreed with us that it would be in our best interests to invest in the new fab. He pointed out that Qualcomm already had offered to invest, and that we all thought that investments by major customers only could lead to undesirable complications.

Our ever-growing difficulties with Qualcomm always came back to the same problem: their rigid specifications.

A specification is nothing more than a number. It represents the level of function that is desired. A chip is designed to do numerous things, and each function has a specification.

The number of chips that meet minimum specifications in any batch is called the yield. A batch of newly designed chips is apt to yield only 30-40 percent that work to expectation. The rest have to be tossed out. Few businesses can survive throwing out two-thirds of their products.

Engineers have to go back into the chip and make changes that will bring up the yield. This is a gradual process, usually requiring many months and presenting occasional pitfalls. A change that increases one function may lower it in another.

Increasing the yield sometimes is just a matter of tinkering. It also may require extensive redesign. With time, the yield usually can be brought up to 95 percent, or so. We always want it to be at least 90 percent. A perfect yield is impossible.

Every chip has to be individually tested to make certain that it works as expected. This is done with automated equipment, and chips that don't meet performance specifications are eliminated.

Since variations occur in the performance of every chip, the customer and manufacturer set an allowable range of performance for every function. This is called the guard band. It simply assures that slight variations in the chip or the testing system will not automatically eliminate a chip. The customer usually wants a narrow guard band, the manufacturer a wider one.

An individual chip could have 20 functions that have to be tested. The slightest variation in only one of those functions could cause the chip to fail.

Unless we could get the yield of the chips for the CDMA phone up, there was no way that we possibly could meet Qualcomm's pressing needs. Yet Qualcomm set such unrealistically high specifications for all functions of the chips and allowed such a narrow guard band for testing that we couldn't do it.

In some cases, we could have increased the yield considerably with just a little more current. But cell phone batteries were not yet as sophisticated as they are now, and Qualcomm wouldn't budge on concessions that would affect battery life and the length of usage. Thus were easy mends denied.

By mid-summer, Qualcomm was demanding chips to make test phones, and we shipped 200 of each of the four on August 14. In the first week of September, we sent another 250 of each.

Qualcomm was ready to begin production, but a large majority of the chips we were making for Qualcomm—many of them completely serviceable—were going straight to the dumpster instead because they narrowly missed meeting a specification or two. By September's end we had shipped a grand total of only 2,505 of each chip that had fully passed our testing, and Qualcomm was even declaring many of these unacceptable.

By this time, patience on both sides was stretched so thin that we never knew what might erupt. Anger, accusations, and harsh words were common.

During all of this contention, a Qualcomm engineer with whom we'd been working regularly from the beginning, a nice, easy-going fellow, simply disappeared without explanation. Only later did we learn that he'd suffered a nervous breakdown. He would be out of work for a year and a half.

One high-level Qualcomm official would get so upset that we frequently feared that something terrible might happen. Everybody at our place dreaded encounters with this guy. I'll call him the Screamer.

Anything could touch him off. He'd fume. He'd cuss. He'd rage. He'd threaten. A simple telephone conversation with the Screamer was enough to ruin anybody's day. A face-to-face confrontation was an ordeal beyond belief.

When things got really bad, somebody would have to go to San Diego and endure the Screamer's tantrums. I usually was that unfortunate soul.

No conversation would last long before the Screamer would fly into an uncontrollable rage, cussing a blue streak. His face would turn beet red. The veins in his neck and forehead would bulge and throb menacingly. I constantly worried that he either would have a stroke—or literally explode before my eyes.

I'd sit there wondering whether I should make any efforts to save him if he did fall over from a stroke—I'd long since ruled out mouth-to-mouth resuscitation—or whether I should be moving my chair back to keep from ruining my suit in case he actually exploded.

It is my nature to attempt to calm anger, seek reason, make a joke, find solutions, work things out, salve hurt feelings. But with this guy none of that worked.

Sometimes he'd get so furious, start shaking so hard, that he'd suddenly leap from his chair, jump up and down a few times, dash out of the room, and disappear, leaving me to wonder what I should do next. Chase him down? Wait for him to return? Alert the police? Call a mental institution?

In all of the business books I'd read, I'd never come across the cuss, get red, shake, and jump up and down school of management, and I was never quite sure how to deal with these episodes.

I'm reminded of a tale by a Southern storyteller named Jerry Clower. A group of hunters trees a raccoon, or maybe it was a big 'possum. One hunter climbs the tree to try to catch the critter and gets into such a long and ferocious battle that he yells down to his companions, "Go on and shoot up in here among us! One has got to have some relief."

We'd just about reached that point with the Screamer. We needed some relief, but we weren't sure where to yell. Eventually, it would require climbing to the highest echelons.

I should add that we were not unsympathetic to the people at Qualcomm. There was no question that this situation was putting the company into quite a fix. It was completely reliant on these parts and had no backup supplier. Every chip that it didn't get from us was a telephone that it couldn't sell and a customer that couldn't be signed up by the service providers to whom Qualcomm had sold its technology.

Qualcomm already were under heavy criticism for delays, and this was causing more.

Yet we were doing all that we possibly could, and still weren't able to meet the company's demands.

This state of affairs was not only straining us emotionally and physically, it was having a drastic financial effect as well.

We already had made six revisions of one Qualcomm CDMA chip, four on each of the other three, with more likely to come. Each of these revisions was a costly proposition, and we were throwing away most of the chips we produced. We already had spent more than $3.1 million developing these chips, but we had received only $442,000 in payment from Qualcomm, much of that in engineering fees. We weren't about to go broke, but we clearly needed to end this drain until we could get the chips to meet enough specifications to produce an adequate yield to serve Qualcomm's needs and allow us a profit.

On October 2, I wrote to Qualcomm what I hoped would be a conciliatory letter explaining the situation and offering a potential solution.

"I will be less than honest if I did not tell you working with Qualcomm has been and continues to be one of the most difficult challenges of my career," I wrote, "and I am sure many of you feel the same about dealing with us....I believe we can overcome all of the differences in our culture and style and emerge from these developments in a successful way. As the business spokesperson for RF Micro Devices, it is apparent we cannot continue with the same financial arrangements we have had through the initial stages of this relationship and I have a proposal which will carry us forward...."

My proposal was that for the next six months Qualcomm would pay us a flat price for each wafer produced containing the CDMA chips. It could set whatever test limits it wanted. We would test every chip and ship those that failed along with those that passed.

This was an effort to temporarily cut our costs and buy time to get a satisfactory yield from the chips.

The letter's effect was immediate. Both sides realized that something had to be done to resolve our problems and move us beyond the continuing acrimony. A meeting was arranged between executives of both companies for October 9 to deal with the situation. The result was a written mutual agreement with the stipulation that no item in it was open to further negotiation.

We did not get the flat wafer pricing that I proposed, but Qualcomm did make many concessions. It accepted a price increase on the chip that had undergone the most revisions and was returning the least yield. It consented to a $500,000 increase in engineering fees to be paid immediately. It agreed to work with us to resolve test issues to allow a greater yield and to change application circuits so that current parts could be used.

We had been having trouble getting TRW to produce enough wafers to meet our needs, partly because of equipment problems at the foundry, and partly because TRW was booking other business. To meet our orders, we had negotiated with TRW to immediately begin turning out 260 wafers per month for us, providing we issued a blanket booking for six months. We assented to granting 90 percent of this capacity to Qualcomm until it could get the CDMA phone to production levels. Qualcomm, in turn, gave us a six-month purchase order for which it agreed to a monthly allotment of no fewer than 60,000 of each of the four chips, and no more than 80,000.

We seemed to be on the way to a peaceful resolution of our problems—at least for a while.

In June, the North Carolina Council for Entrepreneurial Development in the Research Triangle had declared RF Micro the second fastest growing technology business in the state. As flattering and pleasing as this was, by fall we were keenly aware that our growth actually was being severely hampered.

We had to hire new people to meet the huge surge in business now coming from Qualcomm—and soon from Nokia—but we simply had no place to put them. All of our space was exhausted. Our employees hardly could turn around without bumping into one another, and a morale problem was beginning to surface because of it.

The situation got so bad that I returned from a lengthy business trip in October to discover to my surprise that the marketing department was missing.

"We had to move you," Bill told me. "You're up the hill now."

That was literally the case. The marketing department was now in a unit in an adjoining office park a couple of hundred yards away and up a steep slope. Over the next year, I would maintain a well-worn path up and down that hill as I made my way back and forth to meetings.

Shortly after this move, we made a public announcement that we would be constructing a new headquarters building on Thorndike Road sometime in 1996. The developer was Piedmont Land Company of Kernersville, the builder Samet Construction Company of Greensboro.

We had 62 employees at this point and expected to double that number within two years. Where we would put these people until the new building was completed likely would require the hiring of a staff prestidigitator. We hoped to begin construction in January, but government red tape and bad weather would end up delaying that by months.

Venture capitalists have an exit strategy before they put the first dollar into a company. They want to get back their original investment and turn a handsome profit within the shortest period of time. At the beginning, our investors figured that their best exit would be a buyout by a larger company within five years that would return their money 10 times over. That now had changed.

They still were aiming for a return within five years, but now we had major customers and rapidly increasing revenues. The first quarter of our current fiscal year had been our first with more than $1 million in sales, and our second quarter had seen a 1,240 percent increase in orders over the same period for the previous year. Even though our expenditures were growing rapidly and we had yet to come close to showing a profit, our VCs now had determined that a public stock offering would be a better course.

It soon would be four years since our original group had anted up the first $1.5 million, and we now were putting the finishing touches on what our VCs thought would be the last investment before an initial public offering, or IPO as it is commonly called. We already had begun preliminary talks with investment bankers, and we were shooting for an IPO sometime in the second half of 1996. By that time, we hoped, we would have had at least one profitable quarter.

But for now we still needed money. This was a year that had begun with new investment and would close with more. Late in January, we had corralled two new VCs, Alliance Technology Ventures in Atlanta and SVM Star Ventures in Munich. Each had put up $1 million. Our own VCs had added another $1.5 million to bring total investment in the company to $12.5 million.

But our rapid growth and increased expenses, along with the prospect of having to invest in a new HBT foundry at TRW, had prompted our board to begin putting together another round of investment early in the summer. This one started out to be in the range of $7-$10 million, but by year's end had grown to $11 million.

Two of our original investors, Walter Wilkinson's Kitty Hawk Capital and Ray Rund's Brantley Ventures, chose not to participate. But Qualcomm's business and its promise for the future had given us a tantalizing edge, and we had no shortage of others eager to get in.

Qualcomm itself wanted in for $5 million, as did the venture arm of its investment banker, Alex Brown, but both also wanted board seats in return, a prospect we would not consider.

In the end, our VCs allocated $1.5 million to Qualcomm and allowed NationsBank in Charlotte, which later would merge with Bank of America, to put in $4.5 million. The remainder was parcelled out to current investors.

Total investment in the company was now $23.5 million, and the portion now owned by Bill, Powell, and me, the three people who started it, was down from the 36 percent we held after the initial investment to 7.9 percent, a much smaller portion, but a far bigger and more valuable pie.

Despite our hopes and best intentions on both sides, our tribulations with Qualcomm continued, our shipments still restricted by their specifications.

By November, we had three of the chips for the CDMA phone yielding 50 percent or above, but one was still at 35 percent. The low yield was partly due to quality problems at TRW that should be overcome within a month and would bring one chip to an 80 percent yield, two to 75 percent, and the fourth to 60 percent. Design changes were underway that would raise the yield on one chip to 90 percent, two to 85 percent, and the fourth to 80 percent, but that would take longer.

Still, we had no doubt that we were getting close to bringing these chips into range for a reasonable and profitable yield.

But we had a long way to go on the chips for the PCS phone, and the old patterns of impatience, frustration, and anger were once again emerging. Although we had committed to monthly meetings to resolve problems, they had little effect on hour-to-hour and day-to-day dealings between the two companies, some of which had become quite personal to people on both sides. Harsh words still were spoken. The Screamer had resumed screaming.

Even if we could get the PCS chips to pass all of Qualcomm's rigorous requirements and provide a satisfactory yield, we still had no assurance that we would be able to produce them in the numbers that Qualcomm soon would be demanding.

Despite all of our pleas, TRW still hadn't reached a decision on the new fab that we desperately needed. One of our board members, Erik van der Kaay, knew the president of TRW—both of their corporate headquarters were in Cleveland—and had met with him to express our concerns. But all had been to no avail. TRW hadn't even begun the expansion of the current fab that had been promised.

As 1995 drew to a close, we knew that we were headed straight for a defining crisis—and soon.

For certain, just as we expected, 1995 had been a turning point. We just didn't yet know which way the turn was going to take us.

8
A Gun to Our Heads

January 1996 marked the fifth anniversary of the events that led, a month later, to the founding of RF Micro Devices. Those five years seemed to have passed in a flash.

Despite our setbacks and tribulations, we had made incredible progress, so much, in fact, that it had created the gravest problems we yet had faced.

Appeasing and satisfying Qualcomm remained one of our most challenging trials, but the beginning of the new year brought an unexpected and immensely pleasing development in that battle.

Rich Sulpizio, Qualcomm's chief operating officer, attended our board meeting on January 4. Afterward, Dave and one of our board members, Walter Wilkinson, met privately with Rich. Among the matters they wanted to discuss were the continuing tension between our companies and the failure of communication it had created.

Dave pointed out that one person in particular on the Qualcomm side was responsible for a great deal of the animosity that was keeping the two sides from cooperating fully. If that person could be moved to another position, Dave suggested, the situation would be improved immediately.

"It's done," Walter later remembered Rich responding. "Beginning tomorrow, he will not be on the RFMD account."

Even today, I'm unable to express the amazement, the elation, the sheer bliss I felt on learning that the Screamer was gone from my life, that never again would I have to endure one of his unbearable and embarrassing tantrums—or worry that he might explode and ruin one of my suits. It was almost too good to believe.

The person who replaced the Screamer was Don Schrock. Don recently had left Hughes Electronics to join Qualcomm and would go on to great things there. On my first meeting with Don, I knew that everything had changed for the better. I liked him immediately, and I was confident that he was a person with whom we could work. We still would have differences and problems, but now we could deal with them with reason and civility.

At the heart of our problems was TRW's inability to produce enough wafers to meet the business we were bringing in, and in that regard we had made little progress. After their talk following the board meeting, Rich Sulpizio and Dave got on the phone to TRW and squeezed out a few more wafer starts for Qualcomm in coming weeks. That, however, was the equivalent of sticking a Band-Aid on a gaping wound.

We already were taking other measures to try to make an end-run around TRW. We were redesigning two of Qualcomm's CDMA chips in silicon and a third in gallium arsenide MESFET so that eventually we could place those chips at other foundries. We even had gone back to Rockwell International to see if we could run some of the chips in its HBT process. We were expecting prototype chips shortly, but that, like our earlier efforts with Rockwell, would prove to be an ill-fated attempt.

Even if in time we could move some of the CDMA chips to other foundries, that would do nothing to alleviate our current crisis, which was growing by the day. Our immediate concern was filling the allotment we had promised Qualcomm back in October, but we had to satisfy our other HBT customers, too. And the huge surge in business we were expecting from Nokia clearly was going to be more than we could handle. We were sort of like a guy who'd been crawling through the desert desperately searching for a few drops of water suddenly finding himself engulfed in a tidal wave.

To complicate matters, we had begun getting big orders from two Korean companies, Samsung and Goldstar, which Qualcomm had licensed to build CDMA phones. These companies wanted the same chips we were making for Qualcomm, but they weren't ardent sticklers for Qualcomm's overly strict specifications. They just wanted working chips, and they frequently would send runners to pick them up and bring them back by hand.

Meanwhile, the patience Dave had urged us to have with TRW was running out. TRW still hadn't started the promised expansion of the foundry that would relieve our current situation, still hadn't given us an answer on the new fab that would be the solution to our long-term needs and allow us to grow into the company we aspired to be.

I was on the phone to somebody at TRW in Redondo Beach every day dealing with complications, often several times a day, now and then for hours on end. Usually, I was talking with Bob Van Buskirk, the director of gallium arsenide applications, who became a good friend. Sometimes we'd still be on the phone at midnight trying to iron out one problem or another.

"Jerry, I've gotten to where I can't get to sleep at night unless I've talked with you," Bob joked after I offered apologies for one late night call to his home.

Bob and all the other executives at the Space Park were on our side, struggling to get the expansion started and a new fab built. The delay was at corporate headquarters, and we began pushing for action. Both Nokia and Qualcomm were doing likewise without collaborating with us about it, and we never knew exactly what was going on with them at TRW.

TRW's top management finally reached a decision in late January, and it was not what we had hoped. They would do the promised expansion of the foundry but were willing to invest another $20 million in equipment to manufacture the raw wafers on which the chips are made, if that proved necessary.

But if we wanted a new fab, we would have to build it ourselves, or join with somebody else to do it. TRW would be willing to negotiate a license for its HBT technology for uses at 10 gigahertz and below on the radio spectrum, and it would assist in setting up the fab.

As unhappy as we were about it, TRW's decision was understandable. A new fab of the size we needed would require a tremendous investment, perhaps as much as $100 million in California, where land is pricey and environmental restrictions forbidding. But the initial investment would be just the beginning.

Operating a fab is an expensive proposition. TRW didn't have the commercial products or the customers to keep such a fab running. It would be almost completely dependent on us for that.

Despite our incredible growth, we still were a young and unproven company. If TRW built the fab and we didn't produce the business we expected, it would have no use for the facility, be stuck with that huge investment, and have to lay off lots of people.

Understanding this was little solace, however, to our VCs, for whom this decision was a blow.

RF Micro Devices had been conceived to be forever without a fab. That was part of our appeal to investors. We would design, test, and sell chips, all of which could be done relatively inexpensively, but for the costly part—actually making the chips—we'd turn to existing foundries. Never would we have to risk the enormous sums required to build and operate a fab.

Our investors were opposed in principle to building a fab of our own. Al Paladino and Walter Wilkinson, our original VCs, almost got visibly upset just thinking about it.

Back in May, when we first were trying to get TRW to build a fab, the board had appointed Al and Walter as the committee to help get the job done. At that time, we recognized that we would need to invest in the fab, but Al and Walter wanted to keep the amount as low as possible. Under no circumstances, Walter made clear at the time, could we afford to put up as much as half the cost.

"If we could somehow figure out how to raise the money, it would so change the nature of the company in which we all invested and for which we aspire to succeed," he wrote. "It would not be the same ball game, and I don't know whether some of us would even want to be in the park."

Al had an even better reason than Walter to be wary of fabs. Unlike Walter, he had invested in them. Of the seven other semiconductor companies for which his venture fund had put up money, five had fabs. All but one of those failed. The two companies that had no fabs were successful. To Al, putting money into a fab was like throwing it into a blazing pit.

Bill, Powell, Dave, and I had been discussing for months what we would do if TRW didn't build the fab, and we had reached the inevitable conclusion that we would have to build our own. The question was how to bring the investors around to our way of thinking. We didn't have to worry long about that, because shortly after getting the news from TRW, Nokia stepped in to force the issue.

Johan Lofman was Nokia's manager of new technology sourcing, and we had formed a friendship during our long negotiations to create the partnership between the two companies nearly a year earlier. I always could depend on Johan to be straight with me, and he was on this occasion, as usual.

Nokia was on tight schedules in its planning. The company had to know it could get the parts it needed when it needed them. Our current situation offered only uncertainty. If we wanted to keep our relationship with Nokia, Johan told me, we had until the end of February, little more than four weeks away, to take responsibility for establishing a new HBT fab.

It didn't matter what means we chose to do it. We just had to provide the capacity to meet Nokia's plans.

Nokia was fast on its way to becoming the world's biggest producer of mobile phones, and the critical people there recognized that we made the best power amplifiers in the world for those phones. We had no doubt that Nokia's needs would be ravenous.

The threat from competitors wouldn't allow the company to wait. If we didn't commit to the new fab by the deadline, Nokia would have no choice but to find alternatives.

Here we were, poised to become the big company we'd dreamed of being, and it was all coming down to a single decision. We truly were at a crossroads. We could do nothing and remain a small, struggling business without much of a future. Or we could take a big risk and reap huge rewards.

Obviously, however, there was another possibility. Things might not turn out as we expected, and we could fail at great cost to our investors.

But Bill, Powell, and I never seriously entertained that chance. We had the mind-set of entrepreneurs and just didn't think that way. We knew where we were headed, and we were certain that we were on the verge of getting there.

Unfortunately, our investors lacked our conviction.

"Four years and $23 million into the deal and we're asked to support a new fab," Al fumed in a memo.

If this had such bright prospects, he asked, why didn't TRW, a company with immense resources, snap it up?

To our investors, the risks suddenly had been greatly elevated and the possibility of returns, which so far had been zero, would now be significantly lowered.

Our position came down to a few simple questions: Do we just turn our backs on the hundreds of millions of dollars in business that Nokia is willing to hand us? If we do that, can we remain a viable company? And what chance would our investors have of getting back the $23.5 million they already had put up?

The obvious course for us was to investigate all the possibilities of acquiring a fab and to find out what TRW expected for the use of its technology. Even Al and Walter had to agree to that.

Bill, Powell, Dave, and I were well aware that the short path down which Nokia was pushing us was lined with precipitous obstacles and dangerous pitfalls. We knew, too, that fireworks were inevitable along the way. Yet we had no doubt that this was the route to the company's future, and that the board would have no real choice but to follow it to the end. We were in the position of a woman who was halfway through delivering a baby. We could scream and thrash around—and the investors clearly would do that—but there could be no backing out.

The screaming and thrashing weren't long in developing.

To this point, I had been involved with almost all of the negotiations with TRW and Nokia, but Al and Walter felt this to be so important that they took charge of all the bargaining with TRW, including only Dave from management. The founders were left out, perhaps because the board knew our feelings about a new fab and thought we might be too eager to make concessions to get it. All of us were kept fully informed, however.

The first meeting was held February 6 at our offices on Friendly Road. Al and Walter considered the delegation from TRW to be a second-string team, one without the power to make decisions. In fact, this team was just the bearer of the bad tidings.

That news was as explosive as we had expected it to be. In essence it was this: In exchange for a license to the HBT technology and $15 million, TRW wanted half of our company. The proposition was so preposterous that it practically left Al and Walter sputtering in disbelief and indignation. A second day of talks went nowhere.

I must add here that the founders couldn't avoid a bit of amusement at this point, since our VCs now were in a position similar to the one in which we had found ourselves when they came to our aid. Admittedly, it would have been more amusing if our own futures and fortunes weren't again on the line.

None of us wanted to give up any more of the company than was absolutely necessary, of course, but we also knew that this was about more than money.

To venture capitalists, control is vital. Suddenly, our VCs' control was threatened, and they were going to fight to the end to keep it.

A second round of talks was scheduled for February 11 in San Francisco. I wish I could have sat in on this one, just for the sheer entertainment of it, but, alas, some of us had to be content with dispatches from the distant battlefront.

This time TRW sent in the A Team. It was headed by a seasoned battler named Lou Petroni, who was as hard-headed as any of our VCs. Unlike us, TRW even brought reinforcements. A second team was on standby in an adjoining room, with an open telephone line to corporate headquarters in Cleveland.

Our guys were keenly aware that Nokia's deadline had given us an overwhelming disadvantage. TRW had the technology. We had to have it—and by a certain date. They knew that, and it gave them no cause for benevolence or munificence.

Or, as Al put it so succinctly, "They had a gun to our heads."

As outnumbered and outgunned as our guys were, they still put up a scrappy fight.

Lines in the sand were drawn, and that's exactly what they were called in later reports. Our line was that we would give up no more than 30 percent of the company. They had come down to 40 percent.

Later, Al would say that this was one of the most acrimonious meetings in which he'd ever been involved, and knowing and loving Al, that made it a humdinger for sure.

"We all shouted, screamed, left the room, left the hotel...to get our way," Al and Walter later wrote to other investors.

Eyewitnesses told me it was even better than that, but I won't delve into those unsubstantiated reports.

The result, as we all knew from the beginning, was inevitable. When somebody is holding a gun to your head, literal or not, you usually end up doing what they want.

Both sides came away battered, if not bloodied, but with a complicated and sketchy agreement, subject to many changes in detail, that in the end would give TRW 31.2 percent of the company, with a limit of owning 40 percent, while we got $5 million in cash, a $10 million loan down the line, redeemable in stock, and the technology that would assure our future. TRW also got a warrant to purchase up to one million shares of stock at $10 per share between certain future dates, a warrant they eventually would exercise, increasing their share of the company.

Finagling over the terms of this agreement continued right to the end of February. Intense telephone negotiations began on the 25th and continued each day.

Al and Walter fretted about the control that TRW would be able to exert over the company, and we, too, were concerned that TRW wanted a big say in operational matters, which didn't turn out to be the case. By the afternoon of the 28th, we still didn't have a settled agreement, but TRW supposedly was drafting a letter of intent that would be final. In the meantime, we were fast running out of time.

I spoke by phone almost every day with Johan Lofman. Nokia also wanted a signed co-operation agreement with us and TRW, and we were putting the finishing touches on that. On the afternoon of the 28th, I called Johan and told him about the situation. Leap year had given us an extra day on the deadline.

"Jerry, you tell them that they've got 24 hours to get this thing done," Johan said. "We're serious about that."

I passed along his message to all of our folks, as well as to TRW.

At 4:42 that afternoon, a six-page letter of intent arrived by fax from TRW. Our board still wasn't happy with it, but they had no choice but to vote on it. A meeting by conference call was set for the following morning.

Powell and I had no vote. We had stepped down from the board in July 1993, to make room for new investors. Powell remained secretary to the board, however, and both of us still participated in all board meetings. Bill, Powell, Dave, and I had no doubt about the outcome. We knew the board had no choice but to approve the agreement, and we were immensely pleased about it, though careful not to show it.

The conference call included management, board members, and several other investors. I participated by cell phone from my car. Two votes were to be taken, one to assume responsibility for the fab, the other to approve the agreement with TRW. A fair amount of muttering and grumbling and a little posturing went on before the vote, but there were no nays to either issue, only a single "abstain" from Al on both.

I immediately called Johan with the news. Copies of the signed letter of intent were faxed to Finland, and later that day Nokia sent word that we had met its deadline. It also faxed copies of its five-page co-operation agreement with us and TRW, and these were signed the following day.

The co-operation agreement gave us no guarantees of Nokia's business—another reason for apprehension to our board. We simply got first and last bid on power amplifiers for all of Nokia phones. But for now we were the only company that could offer power amps with the efficiency and reliability that it wanted, and we had numerous custom components already in the works for Nokia. Under this agreement, Nokia also would provide us advance notice on all of its future products, so that we could be fully prepared to make certain that we remained Nokia's primary supplier.

In return, among other things, we and TRW gave Nokia first access to all new technologies developed by the two companies. Another provision of the agreement would prove to have profound consequences for us—and much more quickly than we imagined. We and TRW agreed that Nokia would have first and preferred access to production capacity over the next three years, the term of the agreement.

Our board's reluctant decision to take responsibility for a new fab offered us three possibilities.

We could build a new fab of our own, which we estimated would cost $50-$70 million.

We could take over an existing foundry—either leasing it or buying it outright—and convert it to the gallium arsenide HBT process.

Or we could enter into a joint venture with another company either to build a new fab or adapt a current one.

Al and Walter favored the least costly course, a joint venture, preferably with a company operating a gallium arsenide MESFET foundry that could be converted to our use.

Bill, Powell, Dave, and I thought that a new fab of our own would be the best solution. We could put it where we wanted it, build it to meet our exact needs, and operate it without troublesome entanglements with others. If it succeeded, as we felt certain it would, all profits would be ours. And if for some unforeseen reason it failed, we would have only ourselves to blame.

But before we could make a reasonable choice, we had a lot of investigating to do. Over the coming weeks I accompanied teams inspecting fabs all across the country. One was in North Carolina, in the Research Triangle Park. It had been vacant for five years, but it was far bigger, 175,000 square feet, than we needed, and the owner was reluctant to lease only part of it.

We looked at two in California, one in Oregon, another in Colorado, yet another in Missouri.

An ITT foundry in Roanoke, Virginia, just 100 miles north of Greensboro, offered promise. It currently was supplying MESFET chips to us, but it was small and would have to be greatly expanded.

Another company, Alpha Industries in Massachusetts, which also was making MESFET chips, expressed an interest in working with us, but its fab, too, would require a big expansion.

We had a full report ready for the board meeting on April 16.

Our conclusions were that the chances of finding a current foundry in a suitable location that would fit our exact needs were zip.

We had found nothing that wouldn't require such costly and time-consuming renovations as to be impractical.

Neither had we discovered any company eager to begin negotiations on a joint venture. Most of those with whom we'd spoken were wary of potential costs and of TRW's restrictions on the technology.

Dave bravely presented our recommendation: that we build our own fab.

We proposed doing it in three phases.

In phase one, we would construct a building of 50,000-75,000 square feet, but finish only a portion of it. We would acquire minimal equipment to begin the transfer of technology and provide production of 5,000 wafers the first year. This would take 12-18 months to get into operation and cost about $30 million.

In phase two, we would finish more space, add more equipment and people, and bring production up to 25,000 wafers annually. This could take perhaps 18 months more and cost another $25 million.

Up to this point, TRW would be supplying all the raw wafers, as well as continuing to make chips for us. But in phase three, which would cost another $15 million, we would finish the building, bring in equipment to make the raw wafers ourselves, and complete the technology transfer.

The board's reception was less than enthusiastic. Al and Walter still wanted a joint venture.

This board meeting had its brighter moments when we interrupted it to drive out to Deep River Corporate Park for a groundbreaking ceremony for our new headquarters building.

I had written to Governor Jim Hunt to invite him to speak but received no response. Our featured speaker was Dave Vandervoet, vice president and general manager of the Electronics Systems and Technology Division at TRW's Space Park. Bill, Powell, and I had gotten

117

to know Dave well and liked him a lot.

Most speeches at groundbreaking ceremonies are of the sort that are likely to go in one ear and out the other, but Dave's was so memorable that we talk about it yet.

So fired up did Dave get about HBT technology and the effect it was going to have on wireless communications and the future, thanks to all of us, that he wandered off into talking about how a day was coming when anybody could have a cell phone implanted in a tooth. And he clearly wasn't joking.

This was the first I'd ever heard of this concept. As a marketing person, I couldn't imagine what the demand for such a phone might be (who would abide the required drilling?), or what could have prompted somebody to come up with such a notion. I glanced over at Bill to see if he, too, was amused about this. The look he gave me was assurance that as usual we were on the same wavelength, although neither of us dared show it. Dave, after all, was going to become a member of our board if we ever got this fab deal finished.

Nonetheless, we later got lots of laughs discussing how this tooth phone would work. Surely, it would take a talented tongue to dial it.

Somehow, word must have gotten to Dave about our amusement, although he never said anything about it. But years later, he sent me a clipping of an article in which some scientist was talking about the possibility of dental cell phones. Guess I'll just have to chew on that until I see how it actually works.

Not only were we ignored by the governor for our groundbreaking, but only a couple of news media turned out for the occasion. Few people seemed to be as impressed by our remarkable progress as we were, although we were fast reaching the point that we barely could keep up with it.

Even as the first ceremonial clumps of red dirt were being turned at our building site on Thorndike Road on that optimistic spring day, we already were planning a two-story addition that would more than double the building's size.

Every day that we delayed in getting a fab of our own underway was a day that we fell further behind in being able to meet the ever-burgeoning needs of our customers. Dave, Bill, Powell, and I began pushing for a decision at the board meeting scheduled for May 6.

We had prepared a detailed financial plan for a new fab, along with resolutions that would allow us to proceed and to sign the final documents with TRW.

On May 3, we received a letter from Dave Vandervoet and Lou Petroni saying that TRW believed a new foundry was the best approach and that a joint venture would add significant risk to the technology transfer and the "critical timely implementation of an operational foundry." We faxed this letter to board members with a detailed agenda for the upcoming meeting.

The decision we hoped for was not forthcoming. Al remained adamant that a joint venture was the way to go. And to further complicate matters, he was convinced that the terms of the licensing agreement with TRW were keeping other companies from showing interest in such a partnership. The TRW agreement, he maintained, needed serious revisions.

Although our own law firm had been deeply involved with TRW's lawyers in negotiating the final document and had won many concessions, Al decided to have it examined by a major Boston law firm, which had negotiated technology transfers for IBM. It found the agreement deeply flawed and raised many questions about it, all of them supporting Al's arguments on behalf of a joint venture. Al distributed those to management, board members, and other investors on May 24.

Dave, Bill, Powell, and I were concerned that if the board decided to go back into negotiations with TRW, the whole deal might collapse. Even if it didn't, the process might drag on to the point that the effect would be just as devastating. We simply had to get more production capacity and had to do it quickly, if we wanted to remain in business.

Another meeting of the board by conference call was set for May 28 to try to get a final resolution.

On May 27, Dave sent a memo to board members containing his and Bill's rebuttal to the concerns raised by the law firm Al had consulted. Its responses, Dave noted, were based on the assumption that a joint venture was the most desirable alternative. That might be so if we had a potential partner with whom TRW was willing to share its technology and who had the required facility, equipment, skilled workforce, and the desire to work with us, he pointed out. Unfortunately, after diligent effort, we had found no such candidate.

And even if one presented itself, the risks of a joint venture might still outweigh its benefits.

First, we would be turning over to a partner, at no cost, the technology for which we would be paying an enormous price for an exclusive license. And we would not have control. What if the partner wanted to raise prices, and we didn't? What if our business continued to outgrow production capacity, and the partner didn't want to increase it? We could be right back in the exact position in which we now found ourselves with TRW. And what would be the result if we had a falling out with our partner?

Clearly, the only way to avoid these potential pitfalls was to have our own fab.

Al came to the conference call fully prepared to argue his position, and he did so with gusto. But the other board members had been swayed by the cold reality of the situation and voted to approve the TRW agreement, reject the joint venture, and move ahead with plans to build a fab, although we still had under consideration the possibility of leasing or buying one existing foundry. The founders and others in management could hardly contain our relief and joy.

The TRW documents were signed on Friday, June 7, and that same day TRW's bank wired $5 million into our account.

The first public disclosure of our intentions to build a new fab came when I spoke during breakfast at the local Chamber of Commerce's mid-year meeting at the Greensboro Coliseum on June 12. I had been talking with chamber officials for several months trying to learn what tax breaks and incentives might be available from government and other groups if we did decide to build a plant, and they invited me to talk about our plans.

I described the new fab we hoped to build, told what we would be doing there, and how many jobs it would create—at least 125. I said that we'd like to put the plant in Greensboro, but that no verdict had been reached on that.

Justin Catanoso, a business writer for the Greensboro *News & Record* was there, and after breakfast he cornered me and peppered me with questions. While we were talking, Governor Jim Hunt landed in the coliseum parking lot in his state-supplied helicopter and emerged with his entourage to make his own address to the chamber gathering.

The governor always made a big to-do about his efforts to recruit industry to North Carolina, and I couldn't resist mentioning to Catanoso that I'd invited him to speak at the groundbreaking for our corporate headquarters in April, but we never heard from him. My mischievous nature twisted my arm just enough to force me to add that we might have to factor that into our decision on where to build our fab—in North Carolina, or elsewhere.

Before the governor could begin his speech, Catanoso confronted him, wanting to know why he'd shown no interest in the RF Micro project, of which, it was apparent, the governor knew nothing.

"It has not come to my attention," Catanoso later quoted him. "I don't know who dropped the ball, but I will pick it up today."

I was wandering around the business exhibition when a chamber official tracked me down and said the governor wanted to see me. I returned to the room where the governor was delivering his speech.

After he finished, I was introduced to him and he invited me, Chamber chairman Ralph Shelton, and president Peter Reichard into a nearby room for a chat about our company and its plans.

He was all charm, apologizing for not being informed about our groundbreaking, showing great interest in our success. He invited me and the rest of our management team to the governor's mansion to discuss our new fab in greater depth, offered to do whatever he could to help, and promised that we would be hearing soon from top officials in the state department of commerce who would be at our service.

Before we parted, he even gave me a telephone number that he said would reach him directly. He invited me to call at any time.

Later I did just that. I think it was on a weekend, and I no longer remember why I called, although I'm sure it wouldn't have been just to test him. I reached a person who told me the governor was mowing hay on his farm in Wilson County, but that he'd get a message to him that I'd called. I love tractors, and whether he called back or not, I gained a whole new respect for the governor just knowing he was out riding one.

Not 30 minutes later, though, the phone rang and Governor Hunt was on the line.

"Jerry, how are you?" he said with typical political ebullience, something a marketing person is quick to recognize. "What can I do for you?"

Just goes to show, I guess, that sometimes a little mischief can take you a long way.

Our talks with investment bankers had been ongoing, and we still had hopes for an IPO this year. Most of the bankers saw our agreement with TRW and our decision to build a new fab as pluses. None knew about our deal with Nokia because the confidentiality provision prevented us from revealing it.

An IPO was now more crucial because we would need the money to help finance the new fab and get it operating. In the spring, some of the bankers thought that we might be able to make an offering by September, but that now seemed unlikely.

It was generally conceded that we needed to have at least one profitable quarter for an offering to be successful, and with our rapid growth, we could not yet claim that, although we were hoping for it before the year was out. Al thought that we might get by with a couple of quarters in which we surpassed our forecasts, and achieving that seemed certain.

We had, in fact, surpassed our benchmark plan for fiscal '96 by nearly $300,000. Our sales had been more than $9.5 million, nearly six times the previous year's. Indeed, sales for March alone were

$1.8 million, $200,000 more than for all of fiscal '95.

We had forecast sales of $23 million for fiscal year '97, and July brought the news that we had surpassed our projections for the first quarter by $235,000, bringing in more than $3.6 million. We felt certain that we would beat our second quarter predictions as well.

For our board meeting on July 16, Dave laid out a schedule for the IPO. We would appoint a committee to review the investment bankers at this meeting. The committee would schedule presentations by our top choices to be completed by the end of the month. After making our selection, we would begin the long and complicated process that was required, filing with the Securities and Exchange Commission early in September, and hoping to complete the whole thing by early November.

Unfortunately, this was not to be, for the IPO market was slipping into a slump, and our financial situation being what it was, we were persuaded to postpone until conditions improved.

Meanwhile, our plans for the new fab were moving briskly forward. We had put an option on 6.8 acres adjoining our new headquarters property, had gotten zoning rules changed to allow a foundry there, and had hired a firm to begin designing the facility.

We were entertaining offers from two development companies that were willing to purchase the land and put up a shell building to lease back to us, although neither was willing to finish it to the standards required for a fab at their own expense. Nonetheless, this would reduce the investment we would have to put up, which would immensely please our board.

With TRW's help, we were working to arrange leases for much of the equipment we would require, further cutting our initial investment.

We also had hired a fab manager. Art Geiss came to work on July 12. We had lured him away from Alpha Industries in Massachusetts, one of the companies with whom we'd considered a joint venture. Art had a PhD in solid state physics from Brown University and had worked in an ITT gallium arsenide foundry before becoming fab manager for Alpha six years earlier. He was enthusiastic about the challenge of building a new fab, and, like us and so many of our managers, he was the kind of person who would work 16 hours a day and consider it fun.

Despite all of this planning, no firm decision yet had been reached to build the fab in Greensboro.

One of our minor investors was a cement company in Singapore, and its owners, friends of Al's, had gotten the government to launch a campaign to land the fab in that small, Asian island republic. Al, Walter, and Dave flew to New York in June to meet with representatives of the

Singapore government and came back with the news that they were willing to finance a big portion of the fab's cost, debt free.

The problem was that the HBT technology, which had been developed in large part at government expense, could not be transferred outside the country unless every governmental agency involved agreed. And getting all of those approvals through the myriad intricacies of the federal bureaucracy would take more time than we could spare, more time, perhaps, than any of us had on Earth.

The Singapore group understood our need to move more quickly than their interests would allow, but they remained ready to make a substantial investment in the fab, even if we built it in Greensboro, so long as we agreed to put a second fab, should we need it, in Singapore.

Several nearby states also wanted the fab. We had been approached by commerce officials from Virginia, Alabama, and Georgia who were preparing presentations for us to entertain.

This made part of my job easier, because I was spending a lot of time talking with state and local politicians trying to win tax breaks in the event we went ahead with plans for the fab in Greensboro, where we really wanted it to be. The other offers were handy leverage in these negotiations.

Unlike other Southern states, North Carolina offered no incentives to attract new business, and Governor Hunt, a Democrat, was trying to push a bill through the legislature to allow that. The Democrat-controlled Senate had approved the measure, but it had been considerably weakened by the House, where Republicans were dominant, and the bill was now tied up in a committee that was trying to work out the differences between the two versions.

Late in July, in the midst of the battle, Governor Hunt invited Bill, Dave, Powell, and me to the governor's mansion to discuss our plans and talk about what the state could do for us. Before we could get out of Raleigh, our new fab became the primary example of the kind of business that might be lost if the governor's bill didn't win approval. I don't know if that played a role or not, but the bill did pass shortly afterward. We couldn't have been happier, because we would end up benefitting from it.

For the first time, we were getting a lot of news coverage, and we were hopeful that it would help us win tax breaks from the Greensboro City Council and Guilford County Board of Commissioners. I met privately and publicly with each group, and I thought that we would get help from both, but I was wrong.

The first vote, for $200,000 in tax credits from Guilford County, came on August 1. The county commissioners were a notoriously rancorous bunch. Five voted for it, five against. A tied vote was a losing vote. One commissioner was absent, recovering from heart surgery. If he'd been present, we'd have won.

This loss caused us to take a closer look at the attractive package that had been offered by a delegation from Georgia, which had made the strongest pitch among the other states. They wanted us to put the fab in the outskirts of Atlanta and offered considerable incentives. This had additional appeal because nearby Georgia Tech University had a strong program in radio frequency engineering (Bill's son, now one of our engineers, had graduated there), and that would make recruiting easier. A week after the commissioners' vote, we sent a team to Atlanta to inspect three potential sites, and we began talking with developers there.

The setback from the county commissioners also caused me to redouble my efforts with city officials, who were considering $510,000 in tax credits. I spoke at a public hearing on the matter on August 27, my second appearance before the group, and the following day the council approved the credits, which were to be applied over three years to construction of water and sewer lines and road improvements.

The whole thing took place without much fanfare, and a few days later, with our board's approval, we called the governor, the mayor, the chairman of county commissioners, and the Chamber of Commerce to inform them that we would be building the new fab on property adjoining our still unfinished headquarters building at Deep River Corporate Park. We also invited them to an announcement ceremony at the Embassy Suites Hotel on September 4.

This time, Governor Hunt had been quick to accept the invitation to speak. He went on about what a great example RF Micro Devices was for the whole state, and presented Bill, Powell, Dave, and me with a commemorative platter bearing the state seal.

I felt especially good that day. I was serving as master of ceremonies, and I had invited my dad and his older brother, Howard, to attend. They were sitting down front, and my dad was beaming.

I always had been close with my dad, but remarkably the very products we now were making had made it possible for us to be even closer. Dad lived only a few miles from me in the little green house where I had grown up, a house he designed and built himself. He was alone now, since my mother's death a year and a half earlier, a crushing blow to him.

I had been working so hard and so long since we started the company that I rarely had much time to spend with him anymore. But every night when I wasn't traveling on business, I would leave the office, walk to my car, and no matter the hour, I'd dial his number on my cell phone, and we would talk while I made the 40-minute drive home. We'd discuss what we'd done that day, or talk about family, or current events and politics, which he loved, or the events at Poplar Ridge Friends Church, where he remained very active, and where my wife, Linda, and I were members.

When I was a teenager, just starting to drive, Dad would let me take the family Chevrolet out on weekend nights to prowl with my friends. He knew that I was responsible and wouldn't drive fast or recklessly, but he'd told me many times how he worried until he heard the crunch of the tires in the gravel of the driveway and knew that I was home and safe. Now he knew exactly the moment when I got home, because it was ritual.

"Well, I'm pulling into the driveway," I'd tell him.

"Okay, Son," he'd say. "I'm glad you're home; I'll talk to you tomorrow night."

As I introduced dignitaries and spoke about the future of our company at the ceremony at the Embassy Suites Hotel, I found myself thinking back to that day five years earlier when I first brought my dad to see our little venture. I'd boasted of the worldwide companies with which we'd be competing and bragged of the millions of chips that we'd be selling. And now some of those companies I'd named actually were our customers, and we not only were selling millions of chips, we were getting ready to start making them in Greensboro, right at home, a possibility that hadn't crossed our minds back then.

All I had to do was glance at my dad to see his pride in our achievements and to realize the significant difference the technology that we were producing had made in my own life with those nightly mobile father-son chats. Surely, it was having similar effects for many other people in many other ways. The look on my Dad's face gave me one of the most satisfying moments of my life.

In spite of all of our on-going distractions, we still had to attend to the day-to-day business of designing, making, testing, and shipping chips, and that was becoming more complicated, demanding, and frustrating.

We finally were getting Qualcomm's PCS chips to the point that they were providing yield enough to ship, though not enough to be profitable, and in July, we received a blanket order for $12.4 million of those chips to be shipped over a 10-month period.

Qualcomm asked for 80,000 of each chip in July and wanted us to increase the numbers monthly until October when we would have to begin shipping 800,000 a month, half of which were in HBT.

By this time, TRW had begun knocking out walls in preparation for expanding production space to meet our needs, but that would take months to complete. Even when the new machines came on line we barely would be able to meet the demands. Any delays, breakdowns, or other problems, which are always to be expected, would put us in a real bind.

Hanging over us all the while was a certainty we couldn't ignore: Nokia.

We had begun designing custom power amps for Nokia nearly two years earlier, but Nokia had scrapped all of these chips in August 1995, before they ever came close to production. These were for phones that operated on 4.8 volts, the standard for cell phones at the time.

Nokia switched to 3-volt phones, and we had to start anew with the chips. The change would allow Nokia to make smaller phones with smaller batteries, increase talk time, and leapfrog ahead of the competition, which was part of the reason for all the secrecy.

We would begin getting chips for the new Nokia phones ready for production early next year, and Nokia had told us that initially it would need as many as a million a month. The numbers only would increase with time, just as Qualcomm's were doing.

We were several months away from starting construction of the new fab, and once we did, it still would take a year and a half to begin making chips there. It would be another year before we could produce substantial numbers. We had estimates on future orders from both Nokia and Qualcomm, and they spelled trouble. Even with TRW's increased capacity, by next spring, we no longer could supply both.

We had many meetings to discuss and fret about this without coming up with any viable solutions. We were in one of the most distressing situations any company can find itself. We had such a glut of business, but we were unable to supply the products we had created and sold. We were going to have to drop a major customer, and there was no question which customer that would be.

Our contract with Nokia required us to give first priority to its parts when they were ready for production. Qualcomm had to go. We would fulfill as best we could our current obligations, but we would be unable to accept any future orders until we were certain that we could deliver.

It was my job to be the bearer of these bad tidings, and I deeply dreaded it. I would have preferred to put it off as long as possible, but that would be unfair to Qualcomm. As it turned out, events forced me to quick action.

We had come to the inevitable conclusion that we had to drop Qualcomm in mid-September, and shortly afterward a Qualcomm engineer arrived at our place to correlate some tests. Our secrecy agreement with Nokia had forced us to prohibit people from other companies from our testing labs. This engineer got upset about being denied access and called back to his headquarters to complain about it. He also reported that he'd picked up a rumor that we were about to pull the plug on Qualcomm.

Don Schrock called at 9:40 that night, September 18, and left a message on my answering machine wanting to know what was going on. Don and I trusted one another, and I not only liked him, I admired and respected him greatly. I had to break this news to him face-to-face. This wasn't something that could be done by telephone. I called back the next day and told him that I needed to come and talk with him.

The 24th was the first chance that both of us could fit a meeting into our schedules. Kevin Kelly agreed to join me in San Diego and accompany me to the meeting.

Telling Kevin about this was as hard as breaking the news to Don was going to be. Kevin had worked so hard to land Qualcomm as an account for us, had struggled valiantly to keep things on an even keel through all our adversity. And now he was going to lose this big customer because of our growing pains. The cut in his commissions would be a big blow to his income, but he took it in stride and agreed that we had no other choice.

Don led us into a small conference room to talk, and I laid out everything as openly as I could and made no excuses. I told him that we would fulfill current obligations and do everything possible to smooth the transition to other sources.

I could tell that he was angry, but he maintained his composure and said that he understood. He warned that this likely would end the relationship between the two companies, and I said that I realized that, but I hoped it wouldn't, that I'd like to keep the lines of communication open, and that maybe someday when we were a bigger company we could work together again.

I also told Don how much I appreciated all that he had done to make my life easier in the past nine months, and I meant it from the bottom of my heart. We'd had many problems and differences, but with me, he'd always been fair, reasonable, and, so importantly, civil.

All these years later, Don and I still are friends, and we even can look back at this period and laugh about it. But there was nothing funny about it at the time. It remains the most painful experience of my career. We had strived so diligently to get Qualcomm's business, had been through so much turmoil attempting to keep it, only to have to let it go.

Some in our company still harbor bad feelings toward Qualcomm because of all the anguish they endured, but I'm not one of those.

As messy and unpleasant as the whole experience turned out to be, I think it was one of the best things that ever happened to us as a company.

Qualcomm forced a discipline on us that would have taken years to develop otherwise. It caused us to think about things we'd had no time to consider, made us pay strict attention to details and especially to quality control. It taught us to evaluate suppliers closely and without sentimentality. It gave us credibility with TRW and validity with potential investors. Qualcomm actually positioned us to get Nokia's business and to begin moving toward an IPO far sooner than we might have.

I obviously can't speak for Qualcomm, but the ordeal that this turned out to be for so many on both sides wasn't without beneficial aspects for us. It proved that we could face calamity, work our way through it, and grow stronger because of it.

Some might look back at what happened between the two companies and see only disaster, and I'm sure that there are people at Qualcomm who feel that way. But for us it was a blessing.

Qualcomm became our enabler at a time when we desperately needed one. Before that, we were, in essence, the garage operation that some considered us to be. Qualcomm made us into a real company.

When something big happened in the company, or when our employees were working especially hard and making major accomplishments, we would celebrate with a reward that some claim not to exist: a free lunch. We would hire a caterer, set up a tent with folding tables and chairs in the parking lot behind our original modules, and have a cookout.

We staged just such an occasion at the end of September to celebrate our coming new fab and our remarkable sales. Bill, Powell, and I were sitting together, and we were well into the meal before Dean appeared. It was not unusual for Dean to be late to lunch. He was always strapped to his computer by numbers, and it often was hard for him to break away.

This time, Dean had a stunned look on his boyish face, and we couldn't figure, if you'll pardon the expression, what might be going on.

"Guess what," he said.

"What?" Bill asked.

"I've just worked out the books. It looks like we've made money for the quarter."

"WHAT?!!!" Bill practically bellowed. "Are you sure?"

"That's the way it looks."

We couldn't believe it, of course. After nearly six years deep in the red, we actually were beginning to break into the black?

As it turned out, our second quarter was even better than our first. We had predicted sales of nearly $4.2 million and had exceeded that by more than $1.8 million. We had achieved our first $2 million sales month and our first $1 million week. And we actually were showing a modest operating profit, although we remained in the minus column for the year.

I decorated a satellite dish for our first Christmas in
business in 1991, although we were uncertain
whether the company would survive.

It was a happy day when we received Kitty Hawk Capital's
announcement of their original investment
in the company in February 1992.

Our entire staff in April, 1992, our second year in business. We posed in front of our original module at West Friendly Business Park: (left to right) Bill Pratt, Dean Priddy, Connie Cooke, Jill Howard, Kellie Chong, Merci Danielson, Linda Duckworth, myself, Bob Hicks, Powell Seymour.

Bill's smoking caused me to secretly install a huge exhaust fan in his office.

The Last Thing Our Competition Sees Just Before Being Forced Out Of The Box.

HBT stands for heterojunction bipolar transistor. But to us it stands for HardBall Technology because at RF Micro Devices that's the way we like to play the game. Formerly a far-reaching military technology used to control the cold war, we've taken this technology and perfected it for use in our linear power amplifiers creating the most efficient amplifiers in the wireless communications market. HBT permits us to use only a single power source resulting in saved power and space.

Some of our competition has been saying that HBT is experimental and not ready for the market. You will note, however, that those companies don't have HBT. We have tested it, proven it, and are selling it to major suppliers of wireless equipment. And it's proving itself in the field beyond everyone else's belief. Though we're new in the game, we like to play hardball with the big guys. And when you play hardball, the rule is you have to play to win.

MICRO·DEVICES
It's How You Play To Win.

Manufacturers of Quadrature Modulators • Programmable Attenuators • Linear Power Amplifiers • Low Noise Amplifiers/Mixers • Gain Controlled IF Amplifiers • Quadrature Demodulators

7341-D West Friendly Ave. • Greensboro, NC 27410 • Phone: 910-855-8085 • FAX: 910-299-9809

Our hard ball ad in trade magazines in September 1993, was one of a series that brought a lot of attention to our company and made it appear to be bigger than it was.

We were celebrating our success with AT&T's ill-fated Dragon
Phone when I donned a Santa hat for the company Christmas
luncheon in 1993. The hat became a tradition.

Our staff had grown considerably by spring 1994.

We were growing so fast in 1995 that I returned from a trip to find my office moved into another business complex up the hill from our other modules on Friendly Avenue.

Garvin & Associates

Architect's rendering of our first corporate headquarters, which would relieve our fast-growing staff from extremely cramped conditions.

TRW's Dave Vandervoet spoke of the possibility of dental cell phones at the groundbreaking ceremony for the new corporate headquarters in April 1996.

TRW's Bob Van Buskirk (left) and Dave Vandervoet with Dean (right) at groundbreaking.

Over exertion by the executive staff wasn't a concern for this occasion: (left to right) Bill, Dean, our CEO Dave Norbury, Powell, and me.

The last employee cookout at our original site in August 1996.

Greensboro Mayor Carolyn Allen attended the announcement in September 1996, of our plan to build the world's largest gallium arsenide semiconductor foundry.

We started moving into the new headquarters building in
Deep River Business Park in November 1996.

Our executive staff in 1996. Front row (left to right) Bill, myself, and Powell. Back row: Art Geiss, who had been hired to manage our new foundry, Dave, and Dean.

I loved slipping off to watch construction on our foundry.
This shot was made in March 1997.

The new foundry surely was a topic of discussion at this meeting in
our new board room: (left to right) Dave, Bill, Dean, and Art.

By August 1997, the exterior of the new foundry
was nearing completion.

We attended all the trade shows. This was our booth at the
Microwave Theory and Techniques Society Symposium
in Denver in June 1997.

Finished in record time, the new foundry began
production in June 1998.

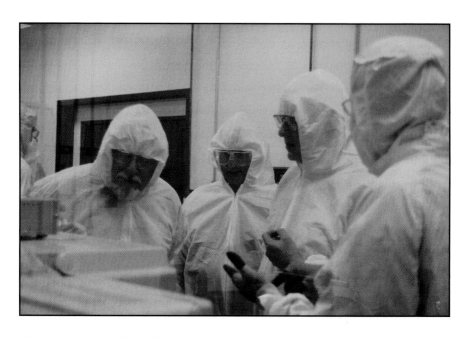

Semiconductor foundries require immaculate conditions and special apparel called bunny suits. Powell, Bill and I donned them to inspect the clean rooms.

By August 1998, we were constructing a big addition to our new corporate headquarters.

I was the only one in a tuxedo and red bow tie at the company Christmas party in 1998. Powell, Bill, and Dean chose less flashy attire.

Governor James B. Hunt attended the groundbreaking ceremony for our third, much larger, foundry in Greensboro in September 1999.

The governor with my wife, Linda, and myself. I was impressed when I called the governor and learned that he was on his farm riding a tractor, one of my great loves.

North Carolina Commerce official Doug Byrd (left) talking with
Bill Shore of Guilford Technical Community College and Dave at
the groundbreaking ceremony.

Dave and Mayor Carolyn Allen led the parade of
spade turners for this groundbreaking.

Bill bet me a bulldozer to a steak dinner that I wouldn't be able to negotiate a new licensing agreement with TRW in 1999. He lost. It made a nice Christmas present.

We barely had moved into our first headquarters building before we had to start planning a new and bigger one. We moved into our present headquarters in January 2000.

Powell (right) and Bob Hicks at the opening celebration
for our new testing facility in June 2000.

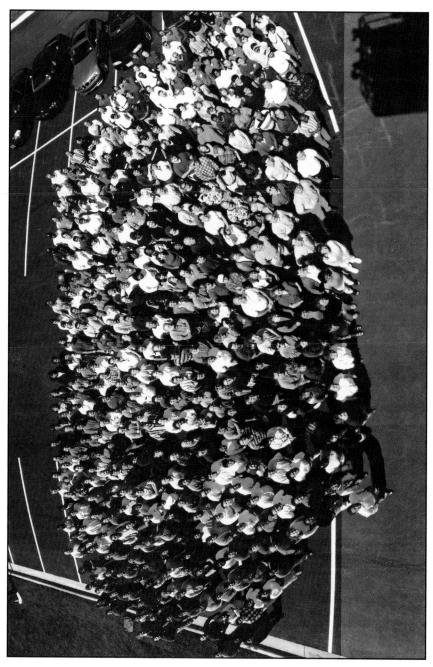

A crane was required to photograph all of our employees in the summer of 2000.

Dave, a master woodworker and a big supporter of
Habitat for Humanity, spoke at the dedication of a house
he and some of our employees helped to build
in the summer of 2002.

Here are some of
our employees
hard at work on
the Habitat for
Humanity house.

RF Micro Devices President Bob Bruggeworth announced our goal of a billion dollars in annual revenue at a company meeting in August 2002. Bob became CEO after Dave's retirement at the end of that year.

Congressman Mel Watts (left) visited our company in April 2003.
He's shaking hands with our vice president for manufacturing
Curt Barratt. Behind Curt is Jeff Shealy, a vice president
at our Charlotte operation.

I was just one of the questionable chefs at an employee cookout in the summer of 2003, at which managers did the grilling and serving.

Following the cookout, our mascot, Wafer Man, took to the field at a Greensboro Bats baseball game for which the company bought most of the seats for our employees and their families.

Congressman Howard Coble visited us in January 2004.

Engineer Jim Christianson (right) shows North Carolina Governor
Mike Easley a semiconductor wafer on a foundry tour in March 2004.

Our employees celebrated the sale of our billionth power amplifier with green t-shirts and a barbecue lunch in August 2004

Bob Bruggeworth spoke of our bright future at a press conference preceding our celebration of selling one billion power amplifiers.

Aerial view of our campus in Deep River Business Park in 2004.

But, from the way things were going, we thought we might even soon overcome those losses. Bookings of new orders were 47 percent above expectations.

Business was so good, in fact, that we raised our sales forecast for fiscal '97. We had been predicting $23 million, but we boosted that to $27 million.

We were moving along rapidly on the new fab as well. We had settled on a developer and a contractor. The developer was expected to be able to complete the purchase of the land by the first of November. The design of the facility was 60 percent complete, and we had determined 80 percent of our equipment needs. We also had hired two more key people to set up the fab and run it.

We now were up to 84 employees, and soon would be adding many more. Because our jobs required highly sophisticated skills, we had to go out-of-state to find most of our people, but we were hoping to change that.

Greensboro is a city with five colleges and universities, but only one of those has an engineering school. That is A&T State University, a traditionally black institution that boasts such well-known graduates as Jesse Jackson and Ron McNair, one of the astronauts who died in the explosion of the Shuttle Challenger in January 1986.

The A&T engineering school had a problem that was just the opposite of ours. Its graduates had little trouble finding jobs, but more than 70 percent had to leave the state to do so, and the university long had been concerned about that.

Most of A&T's engineering graduates would be of little help to us, however, because the school offered no instruction in radio frequency electronics. Early in the summer, we had begun talking to A&T officials about establishing a program that would benefit both of us.

On October 10, we jointly announced a partnership. We would donate testing equipment and chips to set up an RF lab in the engineering school and share our technology and expertise to get students interested in wireless communications. We also would create an intern program to give students hands-on experience in a commercial lab setting.

At the time of the groundbreaking for our new headquarters building, we had expected to be in it by September. Delays due to weather and shortages of building supplies pushed that up to late October, then on into November.

The building was of red Carolina brick, one story, with a flat roof and two wings that joined at a 90-degree angle.

It was situated on a corner. The entrance was at the point where the two wings joined, creating a V-shaped reception area. My office and Bill's were side-by-side in one wing; Dave's, Powell's, and Dean's were in the other. The engineering lab was in our wing, the far bigger testing lab in the other.

We thought the building was beautiful. I went by regularly to check on its progress, and it seemed so wonderfully commodious compared to our six separated and confined modules on Friendly Avenue. As it was nearing completion, I quietly got a crew to install another powerful exhaust fan over Bill's desk so that he could continue to smoke without bothering anybody.

On November 19, the marketing department finally moved into our new quarters. I no longer had to hike half a mile up and down a hill to visit Bill. He was right in the next office, his fan still threatening to suck the pictures off the walls.

It would be after Christmas before all of the departments got settled in the new building and by then it no longer would seem so spacious. Already we were having to alter areas designed for other purposes to make room for the new people we were bringing in.

We let go three of our modules on Friendly Road, but we kept the original three. We still needed the space. The people we were hiring to run the new fab would work there temporarily.

The truth is that after all these years, we still have those first three modules, although they now are used only for storage. Even at the time, if we'd been pressured, we might have admitted to another reason for hanging onto them. At least three of us had a sentimental attachment to the place where the dream that was unfolding before us had been born.

Our new building was not the only reason for celebration as another Christmas neared. We had a perfect present for our board at our December meeting.

Eight months into fiscal year '97, we had sales of more than $16.3 million, $4.4 million ahead of expectations. And for the first time, we were showing an operating profit for the year, $555,412 to be exact.

This news couldn't have come at a better time. The technology market seemed to be improving, and before the year was out, we once again would be moving toward taking the company public.

This had been a year of such incredible developments and profound change in our company's short history that Bill, Powell, and I knew that there probably never would be another to match it. And if another like it did present itself, we weren't sure that we could stand it.

9
The End of the Beginning

A bookshelf in my office strains under the weight of three volumes. They are bound in red vinyl, and each measures nearly a foot in height and 10 inches in depth. All told, they take up nearly a foot of space on the shelf, and they weigh in at a hefty 25 pounds.

These are volumes that require support for reading. And I don't mean sympathetic encouragement, although that would be helpful. I mean physical support.

You could not long hold one of these as you might an ordinary book unless you had the strength of Arnold Schwarzenegger.

Why anybody might actually want to read one is beyond my imagination. I only can advise that any person who might make such a puzzling choice should have a high tolerance for tedium—and perhaps should get a psychological examination at the earliest possible opportunity.

I admit that I have not read any of the three, although I have thumbed through them seeking parts that pertained to me. Those I found are somewhat embarrassing, for they deal with such matters as how much stock I owned and how much money I made in 1997 when these volumes were prepared—something that I always had considered to be a private matter.

But little remains private to those who attempt to take a company public and offer up shares of stock for anybody with sufficient funds to purchase, and that's what these fat volumes are all about. They contain every document involved in the process of turning RF Micro Devices into a publicly traded company.

That process occupied us for the first five months of 1997. We'd made a couple of preliminary stabs at it earlier, but had to pull back each time because the market for IPOs and technology stocks was in decline.

In mid-December of 1996, our investment banking committee, which had been appointed the previous summer, interviewed seven banking firms to choose the underwriters for our IPO. All of these firms had experience with technology companies.

By the end of the month, we had settled on a lead manager. That was Montgomery Securities of San Francisco, which had given us the highest estimated valuation for the company. Two other firms, Oppenheimer & Company of New York, and Hambrecht & Quist also of San Francisco, would serve as co-managers.

This committee, incidentally, was made up of board members Al Paladino, Walter Wilkinson, Erik van der Kaay, Dave Norbury, and Dean Priddy. Dean's presence represented a victory for management.

The chief financial officer is a key figure in taking a company public, and when we had begun seriously working toward that goal, some board members questioned whether Dean's youth and lack of experience with IPOs might be a detriment. Perhaps we should consider hiring a CFO, they suggested.

Both Dave and Bill came immediately to Dean's defense, but were unable to overcome the doubt.

Dave later conferred with nine different investment bankers to get their opinions, and all agreed that IPO experience was helpful but not necessary so long as a person had good technical and communication skills.

He conveyed this to the board members, noting that Dean had the technical skills and that he knew everything about the financial side of the business.

Our management team was extremely unified, and we wanted to keep it that way. Dean was an important part of it. He wanted to see this IPO through, and we wanted him to do it.

Not only would it be unfair to remove him at this point, it would be downright risky, Dave and Bill told the board, and even might delay the IPO. Powell and I stood with them.

The board acquiesced, but insisted that we hire a consultant to coach Dean on presentation for the vital road show in which he and Dave would have to face institutional investors and convince them that our stock was worth buying.

In the fall of 1996, we had two occasions to wonder whether we might need an IPO after all.

We had known for some time that IBM was getting ready to enter the radio frequency market, but we had heard that it was having problems recruiting top engineers. Somebody at IBM got the idea that it might be easier just to buy a company that already was in the business. An IBM official called and asked if we'd be willing to talk with them, and we said sure.

The executives IBM sent to feel us out arrived in October during the international home furnishings market when all the hotels were filled and the restaurants jammed. We took them to a nearby restaurant for lunch and had to wait forever for a table amid crowds that made talking nearly impossible.

I'm not sure whether it was the long wait for lunch, the noisy crowd when we finally got our food, or that IBM just decided it would be cheaper to continue on its course, but this idea quickly fizzled.

One good thing did come from this experience, though. The next day, I got us a membership at the Greensboro City Club so that never again would we have to face the embarrassment of time-consuming waits with growling bellies to take important clients to lunch.

Our second possibility of a buyout began to develop only a few weeks later. It started with a call from my friend Bob Van Buskirk at TRW. Bob told me he'd like to come and talk to our top management about a new idea. I set up a meeting for November 15.

The success of our partnership had led TRW to think that it might be able to find further commercial uses for HBT and create similar partnerships with other entrepreneurial companies that would not be in competition with us. It was considering high-end microelectronics as one possibility, but the small companies in which it was interested were wary of being overwhelmed by a huge defense conglomerate.

TRW was now thinking of buying one or more of these companies, and Bob wanted to know if we'd be interested in taking them in. As part of RF Micro, he thought, they would be in a less bureaucratic setting, more comfortable and compatible, and maybe able to grow faster.

TRW wouldn't just be handing us these companies, of course. Each that we took in would increase TRW's investment, and we wouldn't have to take in but a few before TRW would have controlling interest. Bob knew that this was a testy issue with our board, but we agreed to consider it.

Discussions about this went on for a while, but the question for us eventually came down to this: If we're going to let TRW use this path to gain control, why not just sell them the company outright, let them make it into a separate division of TRW, and fold as many other small companies into it as they want? TRW seemed willing to do this, but we remained so far apart on price that the discussions soon dissolved. Clearly, TRW wasn't ready for a battle, in which we held the advantage.

Instead, we moved ahead as planned with our IPO.

Soon after choosing the investment bankers, we received long lists of information that we had to pull together for them. We had about ten days to do this before the "all hands meeting."

This gathering took place at our new headquarters. It's called the "all hands meeting" because everybody involved is supposed to be present. It included all of the company's top managers, our accountants and lawyers, and all the bankers' representatives and their lawyers. It was an all-day affair where we made lengthy presentations about the company, got to know one another, and discussed the long process that was facing us.

Soon after the "all hands meeting," we were swarmed by SWAT teams sent by the bankers. These were friendly, ambitious, and enthusiastic young people, almost all in their 20s. Their job was to find out everything about the company that would be needed to create a prospectus for potential investors.

These teams had specialties. They streamed in and out of offices and meetings, rummaged through files, and quizzed everybody in sight. It was an intense and distracting period that involved everybody in management, and lots of others as well. All of us spent many hours at it, but a disproportionate amount of the work fell on Dean.

While this was going on, Bill, Powell, and I couldn't help recalling those earliest days when only the three of us had to work night and day to come up with all the information Al and Walter wanted before they would consider that first investment—a huge amount to us at the time, although it now seemed almost insignificant.

A lot of the information these teams were seeking was boilerplate stuff: the company's history; background on its managers, directors, and investors; its technologies and products; its major customers and suppliers; its competitors and financial situation. But such inquisitions also can involve sensitive matters.

I'm told that IPOs have broken down over information that managers simply preferred not to make known.

The information gathering went on for two or three weeks, although it seemed longer. Then another major meeting took place that as far as I know has no special name. This was when all of these SWAT teams got together with representatives of the company and the bankers, the lawyers from both sides, and the accountants to begin hammering out the prospectus. That took place at the offices of Bowne and Company in Charlotte. Bowne, which is headquartered in New York, offers financial printing, among many other services, and would produce the prospectus.

This was not just a single meeting. It was the beginning of a long and difficult ordeal that started on a Monday morning in February. Dave and Dean were our representatives.

The purpose of a prospectus, as Dean so aptly sums it up, is to offer investors information that is "timely, accurate, and properly addresses not only the opportunity but the risks."

At times, however, the matter of what to include can become quite contentious. This, in part, is because the lawyers involved have competing interests.

134

The bankers' lawyers want to highlight the positive stuff that will make it as easy as possible to sell stock. The company's lawyers, who usually prevail in these disputes, want to keep the management and directors from being sued by irate shareholders whose stock value has dropped a nickel and who want to claim that their losses came from being misled by the company. Corporate lawyers apparently spend a great deal of their lives thinking about how to thwart such people.

That's why the "Risk Factors" section usually provides the most intriguing reading in any prospectus. In it can be found any reason that investors might want, or need, to convince them not to buy this company's stock. The potential gloom and doom that is spelled out there is a misery to behold.

Ours would go on for 10 pages of single-spaced tiny type and would include every conceivable calamity that might affect the price of the stock, with the possible exception of the Earth spinning off its axis and disintegrating into cinders in the darkness of the universe.

Come to think of it, that probably was covered under one of the catch-all factors.

The wonder is that anybody who ever read the risk factors in any prospectus would take the chance of buying even a single share of stock, except perhaps as a novelty for cocktail party chatter: "Remember that company that got sucked into the world's biggest sinkhole and just disappeared? I owned a piece of that. But they warned me it might happen."

One major problem did develop during the preparation of the prospectus. We had a written agreement with Nokia to keep its association with us secret. But the law required us to make that public. Although it didn't threaten our relationship, Nokia still didn't like it. But we had no choice.

The writing of the prospectus would take a week of around-the-clock workdays. Dave got to return after the first day, but Dean remained to see it through. He later would recall it as a grueling experience. He usually would work until two or three in the morning, then slip away to sleep a few hours and return to the grind, often finding some of the people he'd left still at work. But by week's end, after lots of rewriting, not to mention muttering and cussing, they had produced the prospectus.

We wanted to file our intentions to make our offering with the Securities and Exchange Commission at the end of January, but the market wasn't doing well and we pulled back. We finally filed on Friday, February 28. With the required documents, we included much of the information from our recently completed prospectus, which was made publicly available in the SEC's files.

Our plan was to sell 2.5 million shares at a proposed price of $11-$13. The stock would be listed on the Nasdaq Stock Exchange, where most of the newer technology stocks are traded. After the filing, we had an indeterminate wait for the SEC to declare it effective. In our case, it would take about a month.

News of the filing broke in the Greensboro News & Record on March 3, and we felt optimistic. But before the SEC approved our proposed sale, our hopes would take a downward turn.

On March 22, the News & Record published an article by business writer Peter Krouse reporting that our IPO was receiving a "cool reception"—not from investors but from people who advise them.

He quoted David Menlow, president of the IPO Financial Network, calling our stock "uninspiring." The big institutional investors who make or break IPOs (mutual funds, pension funds, universities, and the like) no longer had much interest in semiconductors, according to Menlow.

"There's just no sizzle in that market anymore," he said.

Krouse wrote that two other "experts" on stock offerings, Steven Tuen of The IPO Value Monitor, and Manish Shah, publisher of The IPO Maven, were advising their clients to avoid our stock.

Tuen's reason was that we'd lost our biggest customer, Qualcomm, because of "an inability to meet delivery dates and capacity needs," and he was afraid that other customers might abandon us before we could get our new fab up and running. Presumably, he didn't know that we'd had to drop Qualcomm for a far bigger customer, Nokia, and that we were gearing up to begin shipping huge numbers of chips to Nokia.

Shah, who said he actually liked our prospects, predicted that our stock could lose up to 30 percent of its value soon after going public because institutional investors presently were preferring blue chip companies over "small cap" stocks such as ours.

All of these predictions, incidentally, turned out to be flat wrong, as the predictions of stock analysts and market watchers often prove to be. But newspapers rarely go back later to report that the "experts" whom they allowed to give damaging predictions—in this case with no countering opinions—were simply incorrect. And we were powerless to respond because we were forbidden to by law under the "gun jumping" regulations of the Securities Act.

From the time we chose the underwriters, we were not allowed to say anything to anybody outside the company about an IPO, or much of anything else. We received instructions from our lawyers to cease all speeches, advertising, press releases, interviews, financial statements, and any other communications that could in any way be construed as arousing interest in our IPO.

Just over a week after the *News & Record*'s article, we cleared the initial comment stage with the SEC and were ready to proceed with our offering.

The first public presentation of the prospectus is called the "red herring." I didn't have the faintest idea why.

A red herring is a smoked, oily fish with a reddish color, but the phrase also has another definition: "Something that draws attention away from the central issue," according to the *American Heritage Dictionary.*

The dictionary goes on to say that the second definition came from the use of smoked herring to distract hunting dogs from their trails, although it doesn't mention how any prey would be smart enough to catch and smoke a herring, much less use one to fool a dog from chasing it. Not even a fox is that sly.

What I didn't understand is why we would be trying to distract investors who had picked up our scent. It would seem to me that we would be doing whatever we could to lure them on—maybe strewing Godiva chocolates along the trail—instead of using a smelly fish to distract them.

Only later did I discover that it's called a red herring for no better reason than that it has red type on the cover warning that this prospectus is not final and subject to change. And ours, which was made public on May 2, indeed had a change from our initial filing. Because the IPO market seemed to be in decline, and perhaps because of reactions such as those reported in the *News & Record*, the bankers had decided that we should lower the expected stock price to $9-$11.

The *News & Record* reported this on May 8, noting that Steven Tuen of *The IPO Value Monitor* was now "neutral" on our offering but that he expected the stock to sell closer to $9 than $11.

Actually, by this point, things were looking good for the company. We had just completed our third consecutive profitable quarter. Our total sales for fiscal '97 were $28.8 million, even better than our upwardly revised forecast. We also could claim our first profitable year. We had made $1.65 million. We thought that should make our stock more desirable, but it doesn't necessarily pay to think logically when you're dealing with the stock market.

All of the investors in the company had received preferred stock, which in the event of liquidation would be paid ahead of the common stock held by the founders, employees, and to a large extent by TRW. Four classes of preferred stock had been issued, all in varying amounts and at various prices. All of the preferred stock would be converted into common stock for the IPO. The total number of shares of common stock held by investors, founders, and employees before the offering would come to 12,351,441.

Any current stockholders could make some of their shares available at the time of the IPO. It would be the first chance for investors to get a return—or for the founders to see any cash for all of our effort. Any who chose not to offer stock in advance of the sale would face a 180-day lockup after the IPO date, during which they couldn't trade in the stock. Only three current shareholders—Bill, Powell, and myself, chose to offer any. I put up 10,000 shares, Bill 20,000, and Powell 7,000.

A certain amount of stock from the sale would be allocated to the company to be offered to friends, suppliers, and the like. We put out word to our employees that if they had family members who were interested we would include them.

Their faith in the company was far stronger than that of the analysts and market watchers. The response was incredible. Our allotment would be for only 125,000 shares, and we would have needed four times that number to have met all the requests. We took that as a good sign, despite the naysayers.

The next step in the IPO was the most critical: the road show. On May 8, Dave and Dean flew to San Francisco for meetings with the top bankers at Montgomery Securities and Hambrecht & Quist. They put on the presentation they would be giving to investors and made final plans for their trip. Earlier, Dean had gone to San Francisco for two days of intensive training in presentation at Stone Communications. Four days after the San Francisco trip, Dave and Dean were in New York making the same presentation at Oppenheimer, the last rehearsal.

Two days later, on Wednesday, May 14, Dave and Dean flew to Texas, where their first meeting with an investor would take place the following morning in Houston. Over the next two weeks, they would attend 68 such meetings in 16 cities. All but six of those would be what are called one-on-one meetings with individual investors. Each would last about an hour, including time for questions.

The bankers wanted Dave and Dean to get a few presentations under their belts before they got to New York and Boston, where the bulk of the action would take place. That was why the trip began in Houston and Dallas, then moved on to Baltimore and Philadelphia before hitting New York.

At each stop, sales representatives from the banking firms would be waiting and would remain to take orders after Dave and Dean left— "building the book," they call this. Sometimes investors would call Dave or Dean on their cell phones with additional questions as they were heading for the next meeting. The sales reps also would call to let them know whether or not they'd made a sale.

I don't think anybody could question that Dave and Dean were putting on a powerful presentation, because 90 percent of their meetings would produce sales, the largest for 140,000 shares.

Two days were scheduled for the New York meetings, and one of these was at the 21 Club. About 30 investors were expected, but more than twice that showed up.

"Standing room only!" Dave exulted when he called to let us know how things were going. The bankers, he said, were ecstatic.

All of us, in fact, were ecstatic, because the road show was going far better than anybody expected.

After New York, it was apparent that we were going to sell all 2.5 million shares and needed more. The bankers filed a request with the SEC to expand the sale to three million.

In Boston, Dave and Dean endured nine meetings in a single day, and Dean nearly lost his voice. After that, they headed west again, where they hit San Diego, Los Angeles, and San Francisco in another marathon day.

After it was over, Dave and Dean would recall the road show as one of the most exhausting experiences of their lives. It finally ended in Salt Lake City on May 29. Our guys had a chilled bottle of champagne waiting in the limousine, and following that meeting they toasted their incredible success—not to mention their survival.

IPO Day was Tuesday, June 3. The SEC had approved our request to increase the number of shares by 500,000, and the price ended up at $12, just what we'd hoped at the beginning. That day couldn't have been more exciting. Everybody was eager to see RFMD show up on the Nasdaq ticker, and at the office most computers were linked to the Yahoo stock trading site on the Internet.

I couldn't be in on the excitement there. I was manning a booth at a trade show in the convention center in New Orleans, along with some of our reps and others from the marketing department. Everybody was tense about the price at which the stock would begin trading.

When my three children were born, fathers weren't involved at the delivery. You sat in a waiting room with other anxious fathers, nervous and knowing nothing, hoping for somebody to come and give you good news. I had that same feeling at the convention center that morning, but my cell phone now offered opportunity for relief.

I called my assistant, Kathy, at the minute the market opened.

"Any word?" I asked. "Has it shown up yet?"

"No, not yet," she said. Two minutes later I was ringing her back.

A whoop went up in our headquarters building when the stock broke at $15.50, Kathy told me. That was $3.50 over the offering price, a 30 percent increase. I felt like letting out a whoop, too, but I restrained myself considering the setting. Nobody who saw my grin, though, had any doubt that the news was good.

The stock closed that day at $14.87, and to his credit Peter Krouse of the *News & Record* called two of the IPO authorities he had quoted earlier diminishing our offering. Neither acknowledged that they had been wrong. Both credited our success to investors taking a new interest in IPOs. One said that two IPOs by technology-associated companies a couple of weeks earlier—one of them Amazon.com—had given credibility to any IPO with a "high-tech spin." Neither apparently could bring themselves to believe that investors might actually have seen a future for our company.

Krouse also quoted Bill as being "extremely satisfied" with the results of the sale.

"It means there's a lot of other people out there that feel the same way we do," he said.

Obviously, that was so. In little more than a month, RFMD stock would hit a high of $23.75, nearly double its issuing price.

The IPO cost us about $3.5 million, including the fees charged by the investment bankers. From it, we received $33.48 million. Much of it would be used to pay for our new fab.

By early July, we were in one of the busiest periods of our company's history. We had started production on six power amplifiers for Nokia and soon would be shipping 400,000 of each every month, with the number multiplying quickly to 1.4 million. We also had four other power amplifiers under contract with Nokia, with several other parts in planning.

Next door, our new fab was nearing completion far sooner than we had anticipated. Early in the year, we had thought that it would be behind schedule by this point.

Not until the end of January were we finally able to begin clearing the land. Several things had delayed us, one being that the property lay directly beneath a take-off and landing path for the nearby airport. The big jets passed only a few hundred feet overhead. We had to get seismic studies done to ensure that the delicate equipment we would be installing wouldn't be affected by vibrations from jet engines, or from the heavy traffic on nearby I-40.

Another problem was the weather. It turned so foul, and the ground became so saturated that we couldn't get the equipment in to work. The site was in a slight depression, and a lot of dirt had to be moved. A huge ditch had to be dug to accommodate a pre-cast concrete tunnel that would underlay the building. It would be filled with the complex of pipes and power lines necessary to operate the plant.

We finally had to cover the site with huge tarps to keep it dry enough for crews to work. I've had a life-long fascination with heavy equipment, especially earth movers. I'd go over almost every day to watch the huge machines ripping into the red dirt. After we got the foundations poured, the building went up at an incredible pace.

A semiconductor foundry is a complicated structure that requires unique methods of construction. That's because semiconductor chips have to be made in a near sterile environment. This requires "clean rooms," as the work areas are called.

Our building contractor would erect the shell and finish the sections that included offices and laboratories. But the bulk of the building would be made up of clean rooms, and they had to be installed by a company that specialized in that work. Ours was Suitt Construction Company, with headquarters in Greenville, South Carolina.

The creation of a clean room has to be...well, clean. The construction crews leave no debris at the end of a workday, no smudges, no dust. Everything is constantly wiped clean as the work proceeds.

Special paints and finishes that resist static electricity and the collection of particles from the air have to be used. The floor has to be raised and grated so that air can flow through it. Heavy equipment is set on concrete pedestals around which the raised floor is constructed.

A single-story clean room needs to be two stories high. The second level houses the huge, powerful fans and highly efficient filters that constantly cleanse the air of particles.

Unseen debris from the air—a speck of dust, a flake of skin, an eye lash—can cause great damage to a wafer filled with chips. Under magnification, a speck of dust can look like a beach ball imbedded in a chip.

Different steps in the making of microchips require different levels of air purification. Clean rooms are classified by the amount of particles that are removed—the lower the class number, the cleaner the air.

The cleaned air flows out through the grated floor, and new air comes in from outside. The air inside a clean room is completely replaced every few minutes.

Static electricity can be deadly to chip making. It can destroy a wafer filled with thousands of dollars worth of chips, and extensive steps have to be taken to control it. Clean room employees must wear anti-static straps, bracelets with a wire that must be plugged into a grounding plug at each station where employees work before they can touch anything.

A steam mist is used to maintain humidity at certain levels to reduce static electricity. Ionizers, which alter the electrons of the atoms in the air to make it slightly conductive, also reducing the creation of static electricity, hang at regular intervals from the ceiling.

The water used in the manufacturing process has to be free of all contaminants and deionized as well, and huge pumping and filtering systems are required to purify it.

People who work in clean rooms first enter a room where they receive air showers that remove particles from their bodies and clothing. They move on to a changing room, where they don what are called "bunny suits" in the trade, white, sterile protective gear that covers their entire bodies except for their eyes—coveralls, hoods, masks, booties. If they leave the work area to go to the rest room, or for any other purpose, they must dispose of their outfits and put on new ones when they return.

Needless to say, the operation of a clean room is extraordinarily expensive. The power bills alone are astronomical.

To me, there always has been an appealing, other-worldly feeling about clean rooms. The employees in their bunny suits, the futuristic equipment with its soft whirrings and flickering lights, the ionizing tubes, the sealed observation windows, all make it seem like something from a science-fiction movie.

Remarkably, by late August, our building was completed, our clean rooms finished, on budget and far ahead of schedule. The clean rooms were certified by an industry quality assurance group on September 6, and the following day we began moving in equipment.

Our board met on the 9th, and we took them on a tour of this impressive structure that some of them had opposed. Nothing was said about it, but I had no doubt that all of us were feeling the intense pressure of making certain that this thing succeeded.

Already we were moving ahead with future phases of the fab. In the beginning phase, we still would be dependent on TRW for the wafers on which the chips are made. Gallium arsenide wafers are created by a process called molecular beam epitaxy in machines that cost millions of dollars each. We had ordered our first MBE machine, which we hoped to install within a year, and at the board meeting we had discussed ordering a second.

We were moving ahead in other areas as well. The architectural plans for the addition onto our new headquarters were complete, and we expected to break ground at the first of the year. This addition would be bigger than our original building, two stories with 34,600 square feet. It would include a state-of-the-art testing lab of 11,000 square feet, office space for an additional 140 people, bigger conference rooms, and a cafeteria that would serve hot meals and seat 90. We also were nearing closing a deal on 30 acres across the road, where we hoped to build a far bigger and grander headquarters building. We were preparing to hire a civil engineer to begin land studies and cost estimates.

By the end of October, we had 200 employees, nearly 50 of them specialists recently hired to operate the fab, and all were now working there with teams that TRW had sent to help get the plant going. The equipment installation was nearly complete, three weeks ahead of schedule, and already we were beginning some of the critical preliminary steps of the technology transfer. Thirty-two more people were getting ready to move from the headquarters building into new offices in the fab. And we recently had hired a manager for the wafer-making operation, which still was at least a year away.

We now were shipping chips to Nokia as fast as we could turn them out, and bookings for new orders were setting all-time records. Another Korean company, Hyundai, had become a big customer, and for the first time Motorola had designed an RFMD power amp into one of its mobile phones. We even had gotten back the very first customer who had paid us money, then dumped us, Nippondenso, now called Denso. The company had ordered three parts for a new CDMA phone to be used in Japan.

One night near the end of October as I was working late, I got restless and decided to walk over to see what was happening at the fab. It was hard to believe that just eight months earlier, I had watched bulldozers tearing into the red dirt, and now on this spot stood an impressive building, with lights aglow and people coming and going.

Every fab I'd ever seen was a drab, practical structure of pre-cast concrete, but we had designed ours to complement the headquarters and built it of red brick. It was an attractive building.

When I went inside and walked down the long corridors peering through the viewing windows into the clean rooms that had been finished, watching the people in their bunny suits hard at work, a feeling of awe at what we had accomplished in so short a time suddenly overwhelmed me.

But I also was struck with another realization, and for the first time I truly understood why Al, in particular, and Walter, and perhaps some of the other investors had been so reluctant to create this astounding facility.

A fab, in its complex reality, is not just a marvel of human ingenuity. It also can be a ravenous monster. It had to be fed 24 hours a day, seven days a week without letup. We had staked our company on this monster, and if something went wrong and we couldn't get and retain the customers to keep it fed, it would devour us. The weight of that possibility suddenly hit me like the world on Atlas' shoulders.

Bill was still in his office when I returned. I went in, sat down beside his desk, and told him about my walk through the fab and the feelings that had boiled up so unexpectedly. I realized it was uncharacteristic of me to be having such feelings, because from childhood I had learned that hope and optimism inevitably pay dividends over doubt and pessimism, but I always could express whatever I was thinking with Bill.

"This could be the beginning of the end," I told him.

Bill smiled and echoed Winston Churchill in 1942.

"I don't know if it's the beginning of the end," he said, "but for sure it's the end of the beginning."

10
Wall Street's Darling

In expressing their early opposition to building the fab, some of our board members said that it would completely change the nature of the company, and they were right. But they had been speaking only in financial terms—the amount of investment and the risk involved. Our fab actually changed us in more profound ways.

Suddenly, the scale was completely different. In the past, we had bickered and worried over almost every expenditure, had taken great pride in each substantial sale, and had suffered over every setback, big and small. But after we committed to the fab, with all of its demands and its big and burdensome expenses, matters that once had seemed huge became almost insignificant in comparison.

We had reached an entirely different level. Nothing we had done before ever would be adequate again. Everything had to be bigger and better, and we had no choice about it. We always had operated with a sense of urgency, but now that urgency became far more intense. We couldn't slack off for a minute.

We knew that a fab could be an incredible money maker, but voracious devourers of cash that fabs are, we were keenly aware that they had brought about the ruin of some companies. We, however, had one big advantage that other companies didn't: Nokia.

Nokia was growing so fast and its projections were so great that we had no question that this fab would pay its way—so long as we were able to please Nokia. Dave made clear our intention to do that in a long-range plan that he presented to the board in March 1998. Foremost among our many goals: "Keep Nokia at all costs."

Our sense of urgency and our need to accommodate Nokia had caused us to push hard to bring the fab online ahead of schedule, and we did it. At the beginning of June, only 16 months after the site clearing began, we started production six months earlier than anticipated, a magnificent accomplishment for everybody involved, particularly for Art Geiss and his crew.

We had 63 employees in the fab and were adding others as fast as we could. We were anticipating turning out at least 100 four-inch wafers a week in the beginning, each containing thousands more chips than the three-inch wafers we got from TRW.

We had no time for ceremonies to mark this occasion, auspicious though it was, but we did send out a news release. We made big news locally and got onto the national business wires. Some of the stories noted that we now owned and were operating the world's largest gallium arsenide HBT foundry. Not bad for a company that only a few years earlier consisted of three guys in a small module on Friendly Avenue, three computers, a single-page copy machine, and a few pieces of testing equipment.

In the year that had passed since our IPO, our stock had ranged in price from just under $10 to almost $16, with the exception of that brief spurt into the $20s shortly after it was issued. On March 13, Dale Pfau, an analyst for CIBC Oppenheimer, noted that research indicated that market trends favored gallium arsenide HBT for power amps in cell phone handsets.

"RF Micro Devices is the clear winner in this trend," he wrote, "and although the battle will be played out over the next year or so, if we are correct, our $17 to $19 price target will be far too conservative."

That gave our stock a temporary boost, but on the day of our news release about the new fab, it was trading at about $12.50, $2.37 less than it had closed on the day of the IPO. The news caused it to rise 87.5 cents. If that was a show of confidence, it wouldn't hold. By the end of the month the stock price dropped to $10.88.

That took nothing from the luster and importance of our fab, however. I loved going there and watching the employees and the vital work they were doing, loved taking people there to show off the place. The fab gave our company a whole different feel, a new and grander sense of ourselves.

And it gave us a credibility with our customers that we couldn't have gotten any other way. They knew now that we were committed, that we had substance, that we could produce without dependence on others, and that we were a force to be reckoned with.

In mid-July we began shipping from the new fab, and a week later we announced that revenue for the first quarter of fiscal 1999 was up 87 percent over the same quarter for the previous year. Both announcements sent our stock upward, and over the next two months, it would trade, with the exception of one brief dip, in the $16 range, although the market itself would be in a slump brought about by worries over economic turmoil in Asia.

North Carolina was undergoing a drought that summer, and Greensboro's already limited water supplies were getting low; some areas of the reservoirs turned into cracked and drying mud flats. The water we were getting from the city was so filled with organic matter that it kept clogging our filters. We never had any production delays because of it, but it was a constant aggravation that cost us a lot of time and money. It later would become a matter of such serious concern that we would have to dig a series of deep wells, the first in 1999, to assure constant supply.

As summer neared its end, on Monday, September 14, TRW redeemed its warrant for a million shares of stock at $10 a share. That gave them a quick and tidy profit, since the stock closed that day at $16.50. TRW now owned more than 5.6 million shares and increased its ownership in the company to 32.7 percent.

The following day, we announced that revenue for the second quarter would be higher than expected. Analysts had been estimating $25.6 million. We predicted $28–$30 million. Actually, sales would be much higher, $31.4 million, up 243 percent over the same quarter for the previous year, with a profit of $2.36 million, although that announcement wouldn't come until October 20. But the conservative prediction itself caused the stock to jump to $20, although it would fall back quickly.

By this point, we had 80 employees in the new fab, 300 in the company, and production was gradually growing. Even with the new fab, we still were struggling to meet Nokia's ever increasing needs. That caused us to push ahead with expanding the fab long before schedule.

In our original plan, the fab would be built in three phases over a period of four and a half years. But the urgency to rapidly increase production had prompted us to combine phases two and three and cut the time by more than half. By early September, we were installing our first molecular beam epitaxy machine to produce the raw wafers on which the chips are made, originally part of phase three. Two more were on the way, and the board would approve a fourth at its meeting on October 27.

The $10 million we got from the exercise of TRW's warrant would help to pay for the expansion, but Dean, who had become chief financial officer after the IPO, announced that we also would be borrowing $15–$20 million for that purpose.

In addition to the fab expansion, we had set up a manual, prototype packaging line at our original units on Friendly Road to add the connectors to the new chips we were designing. This would allow us to get new parts to market much faster.

• • •

After the crash of Japan's stock market that began late in 1989 and sent the country's economy into a decade-long tailspin, the economies of other Asian nations boomed, fueled by Western investment and cheap exports. That included South Korea, where cell phones were selling like mad and we had several big customers who were giving us millions in business annually.

The great wealth that was amassing in Asia by the mid-1990s brought corruption and reckless investment that reduced the return to investors, causing many from the West to pull out of those markets. Asian investors, too, began moving money out of the region.

The huge outflow put a strain on the currencies of those countries, and after several failed attempts to maintain the value of the baht, Thailand allowed it to trade freely on the currency market against the U.S. dollar on July 2, 1997. The baht fell 20 percent that day and would continue to drop, setting off a chain reaction that brought an unprecedented collapse of currencies and stock markets throughout Asia in the coming months.

The crash wasn't without effect in the United States. On October 27, 1997, the Dow Jones Industrial Average fell 554 points, the greatest one-day drop in history to that point. Technology stocks were especially hard hit because many high-tech companies did huge amounts of business in Asia. Some technology stocks would remain repressed for two years.

Although our orders from Korea came to a temporary halt, cutting our sales by several million dollars in months to come and causing our stock to dip, we didn't take the hit that some technology companies did. Our growing business from Nokia and other companies covered the losses, and by the summer of 1998, months before stock analysts would begin declaring that the Asian markets finally had hit rock bottom, we were seeing our Korean orders coming back strong.

By October 1998, the collapse of the Asian markets was still rippling through economies worldwide, but especially so in the United States. Securities firms, which had seen their own profits and stock prices fall drastically, were shedding employees by the thousands.

But, ironically, for most of our economy, Asia's miseries were turning into our good fortune. The price of oil had fallen, as had prices of other commodities. Inflation had been stopped dead. Interest rates had dropped, and our economy was headed into a boom the likes of which rarely had been seen.

After climbing briefly to $20 in September, our stock had dropped back to $16 early in October. But then *Inc.* magazine named us one of the 500 fastest growing companies in the nation. And the public announcement on October 20 of our huge increase in second quarter sales and profits not only sent our stock price climbing but caused analysts to give us "buy" recommendations. Charlie Glavin of Credit Suisse First Boston said that the only reason he didn't give us a "strong buy" was because of investor jitters over the market turmoil. He called us a "top-tier company from an investment perspective."

Our stock continued going up. On the day before Thanksgiving, it closed at $28.69, a new high.

By December, we had three MBE machines churning out the raw wafers that our new fab craved, each in separate bays in the clean room area. These machines had to be regularly broken down for cleaning and servicing, and because hazardous materials, including arsenic, were involved, extreme safety measures had to be taken.

One of the new machines was undergoing this process on Friday, December 4. The crew handling the job finished shortly before 7 p.m. and took a break. The bay door closed automatically behind them. When the crew returned to haul away the waste, they found the room filled with smoke.

The entire fab was immediately evacuated and the fire department called. After the firemen arrived, while discussions were underway about how to handle the situation, the sprinkler system activated in the bay where the MBE machine was situated, extinguishing the source of the smoke. Nobody ever saw a blaze.

Later investigation would show that materials in a waste barrel had spontaneously combusted, creating a smoldering mass that melted a plastic vacuuming hose, producing heat enough to set off the sprinklers.

Our employees had been properly following procedures, but obviously those procedures were flawed. It would take two years of research to determine what caused the cleaning materials to ignite, but we immediately changed procedures to make certain that no such incident ever happened again.

A little smoke had seeped into other areas of the fab, but that night we called in a company that dealt in the cleanup of hazardous materials. Their crews sealed off the involved area, inspected other areas, and did minor cleaning of smoke stains. By the following morning, we were able to reopen the fab.

The MBE machine that was undergoing cleaning eventually would be declared a total loss because of water damage. It and the bay in which it sat would be dismantled and hauled away over a period of weeks.

The loss of this expensive machine was a setback at a time when we were struggling to increase production, but we still could get wafers from TRW to fill in the slack, and we had another MBE machine on the way. We were relieved and grateful that no employees were injured or subjected to potential harm.

The fire aside, December turned out to be an incredible month for us. On the last day of November, another brokerage firm, Dain Rauscher Wessels, began covering our stock, opening with a "buy" recommendation and a 12-month target price of $35 per share. Two days later, on the day before the fire, Charlie Glavin of Credit Suisse First Boston raised our rating to "strong buy." On the day of the fire, our stock jumped $4 to close at $37.25.

The following week we reported that revenue for the third quarter would be much higher than expected, $38–$39 million, beating analysts predictions by at least $5 million. Profit, we thought, would be between $4.7 and $5.1 million. As it turned out, we had been conservative. The actual figures would prove to be $41.5 million in sales with $5.6 million profit.

The next day our stock surged 27 percent to close at $46, an increase of 322 percent from our low of $10.88 in June. That day Credit Suisse First Boston raised our 12-month target price from $40 to $55.

The national business press didn't fail to take notice either. The following Monday morning, Dave was interviewed on CNBC's whistles-and-bells pre-market-opening show *Squawk Box*. Several of the company's officers, myself included, seeing the first real chance to profit since the IPO, had sold stock when it had risen above $20 in late November. The interviewer asked Dave why insiders recently had been "dumping" stock in the $20 range.

Dave grinned. "They're kicking themselves in the tail for doing that," he said.

Shortly before Christmas, we filed notice with the SEC of our intention to make a secondary stock offering in January of two million shares, including 250,000 held by TRW. With the stock price still rising, the time was right, and we needed money because we had big plans for the coming year.

Indeed, although we couldn't know it at the time, 1999 would prove to be the most spectacular year in the company's history. Even now, I find it hard to believe all that happened.

Ten days into the new year, a huge, boldfaced headline topped the entire width of the *News & Record*'s business page: "Triad ticker: Stocks are soaring." Among 20 companies in the area whose stocks were faring well, we were number one. The story, written by Doug Campbell, began this way:

"RF Micro Devices suffered no sophomore slump. The Greensboro-based microchip maker was Wall Street's darling in 1998, its second as a publicly traded firm, rolling to a mammoth 276 percent annual return."

This was a great moment. It marked the first time that we ever had seen our company described as "Wall Street's darling."

I guess it's good to be anybody's darling, but to be considered Wall Street's is unbelievably satisfying. It has a certain cachet about it. Or maybe it's just a certain ring. Ka-ching. Ka-ching.

Our board meeting on January 26 was one of immense consequence for the company. Well before we got our new fab up and running, we were aware that it was going to be far from adequate even if we did rush it into full service.

Cell phones only had begun to penetrate the world market. Analysts were predicting that sales would grow by as much as 40 percent annually for the next decade. Some were saying that, within a few years, sales could reach a billion a year. Our current orders and Nokia's projections told us that we had to have more production capacity—and sooner than we had thought possible.

At this meeting, without objections, the board approved two new fabs. The first, which would be dubbed Fab Two, would accommodate only MBE machines to make the raw wafers for our chip production lines. We planned to put this fab in a leased 136,000-square-foot shell building in a business park off Gallimore Dairy Road, just over a mile from our current facilities.

The MBE machines in our current fab, which from now on would be referred to as Fab One, would be moved to this location, thus freeing up more space for chip production. We would build clean rooms in only 48,000 square feet of Fab Two at the beginning, and planned to have it in service no later than September 1.

Fab Three would be designed to give us seven times the capacity we would have when Fab One reached full production of more than 30,000 wafers annually within the coming year (actually it eventually would produce twice that). We estimated the cost of Fab Three to be $200 million, and we hoped to build it on a parcel of land near Fab One, but that wouldn't be decided until later.

Although we still were a month away from moving into the new 50,000-square-foot addition to our current headquarters building, authorization also was granted to begin construction in April of a new 100,000-square-foot, two-story headquarters on the 30 acres we had bought across the street from our current facilities. The cost would be about $8 million.

To help get production moving faster in our current fab, the board approved adding a fourth shift so that we could operate 24 hours a day, seven days a week.

This board meeting assured that we truly were on the way to becoming the big company we had intended to be from the beginning.

Our stock continued to rise. Our secondary stock offering at the end of January went at $61.45 per share, bringing us $133 million after fees. By mid-February, when we announced plans for our new headquarters, the stock was trading in the $70 range. News stories reported that the new building would accommodate 500 employees, which spoke to our expectations, since we had only 400 at the time.

On the last day of February, the stock rose above $84 before settling back to close at $79. The next day, it was trading in the $80s again, too high, we thought, to entice some buyers.

At our board meeting on March 2, the only decision that would make the news was the one to issue a two-for-one stock split effective March 17, which would cut the price of the stock in half, and, we hoped, make it more appealing to more investors. At the same time, *Money* magazine placed us at number seven on its list of best-performing stocks over the past year.

A day after the stock split was announced, the *Business Journal of the Triad* revealed that we had leased the building on Gallimore Dairy Road for our second fab and had hired a company to renovate it with clean rooms at a cost of $8 million. The work was to be finished in May. This news, combined with the coming stock split, would send the stock price almost to $104 on the day the split took place, nearly a tenfold increase in only nine months.

As strange as it may seem, our paramount concern at this time was not our stock price or production capabilities, but our long-ongoing effort to diversify our customer and product bases, because that was where the company's future really lay.

From the beginning, we knew that gallium arsenide HBT was a tiny niche in the semiconductor business. We never thought of ourselves as a niche company and didn't intend to be one.

Gallium arsenide HBT was a great base for power amplifiers—and they by far made up the major part of our business—but we also knew that this technology wouldn't last forever. Something eventually would replace it, and whatever that might be, we wanted to be there before anybody else.

We also realized that the semiconductor industry was built on silicon. It was fundamental, simpler, and cheaper than other technologies, and it was where the real money was to be made.

Every part in a cell phone could be done well enough in silicon, with the exception of the power amp, and in any given phone, 90 percent or more of the parts were based in silicon.

We constantly were working to increase our silicon business, which now amounted to only about 12–13 percent of sales. Gallium arsenide HBT, on the other hand, usually came in at 85 percent, or so, with gallium arsenide MESFET making up the remainder. Although we had no intentions of building a silicon foundry, which would require an investment far beyond anything we had in the works, or could hope to fund at this point, we wanted to get silicon up to 50 percent or more of our sales.

One technology that was considered to be "hot" at this time was silicon germanium, a bonding of the two elements first used separately in integrated circuits. Silicon germanium was the closest silicon alternative to gallium arsenide HBT, faster and more efficient than silicon alone. It functioned at high frequencies and was less expensive to produce than gallium arsenide. We long had been considering it for CDMA parts and had determined that it would be a great technology for low noise amplifiers, upconverters, downconverters, and possibly even for power amplifiers.

In March, one analyst claimed that we already had silicon germanium, but we didn't. IBM did. It was our silicon technology partner, and had been making all of our silicon parts for years.

But when we told IBM that we wanted to start producing silicon germanium components, IBM balked. It was now in the radio frequency chip business itself and wanted to keep that technology.

When it became clear that IBM had no intention of working with us on this, I began devising a strategy to overcome the problem, but that would take a while to pull off.

Silicon germanium was just one part of our effort to diversify, though. We also were driven by a more obvious need.

We loved Nokia and were gratified to have it as our biggest customer. In any given quarter, Nokia accounted for anywhere from 60–85 percent of our sales. We also were all too aware of how risky it was to be so dependent on a single customer.

If we lost Nokia at this stage, we might not be able to recover. We knew that many people, including competitors and stock analysts, kept a vigilant eye on our relationship, searching for signs of strain or dissatisfaction, and at the first hint of any such trouble, our stock could sink like a rock. Our challenge was to keep Nokia and have its business with us continue to grow, while we worked doggedly to design new products and bring in new customers that would gradually reduce Nokia's percentage of our sales and cut our dependence on the company. Nokia, too, wanted this. It would have been far more comfortable if they accounted for only about a third of our revenues.

Much of my time was devoted to our goal of expanding silicon and gaining new customers. I was always coming up with one scheme or another to make it happen, and to some extent, we were accomplishing that, but not nearly enough at this point. And that only added to our sense of urgency.

Nonetheless, our customer lists and our products continued to grow. Our 1999 catalog was a fat 1,260 pages, compared to a heavily padded 10 for our first one seven years earlier.

On Sunday, March 14, three days before our scheduled stock split, the *News & Record* published a lengthy story about the company written by Peter Krouse. It was accompanied by lots of photos and a chart showing the amazing surge in our stock price. Krouse wrote that we had "staged a virtuoso performance" since our IPO and were getting "rave reviews."

Much ado was made about the company's success and how we had achieved it in a relatively short period. A fund manager described us as being in an "explosive stage of growth." Bill called the past year "phenomenal" and noted that our goal was to be "the Intel of the wireless world."

But the article went on to talk about our personal "new found wealth," noting that Bill recently had bought a house on the ocean front at Wrightsville Beach. The most extravagant expenditure that could be pinned on me was my new John Deere tractor, a long-held dream of which I was quite proud.

Krouse did point out that we didn't like to talk about such matters.

"It's just unseemly," he correctly quoted me. "We didn't get into this to get wealthy."

I continue to feel exactly that way, and so do Bill and Powell.

One aspect of this article did amuse me. In it, Bill was described as "the sober-minded Pratt"; Powell was "the stocky Seymour," while I was "the affable Neal." Don't ever let anybody tell you that affability doesn't pay.

Getting good and experienced engineers in our field was growing more difficult, and sometimes when we did find them, they, like we, were settled in an area, and for the sake of their families, or other reasons, didn't want to leave.

Early in the year, Bill heard from an engineer at a British company that made analog signal parts in Scotts Valley, California. The plant was going to shut down, and he and several engineers were going to be laid off. They were interested in working for us but wanted to stay together and in California. Would we consider hiring them as a group and opening a branch office in Scotts Valley?

We flew them in, decided they were good people, and quickly brought them into the company. On April 5, we opened our first outlying design center in a rented facility in Scotts Valley, a half-hour drive from San Jose, giving us a foothold in Silicon Valley. It employed 15 and included marketing and customer support people in addition to the engineers.

"This opportunity fell into our lap," Dave told reporter Eleni Chamis of the *Winston-Salem Journal*. "It's the kind of thing we need to do to find more engineers. We're turning business away every day because we don't have enough technical people."

Word of this would spread through other companies around the country, causing other groups of engineers to reach out to us. After Rockwell International bought Collins Radio in Cedar Rapids, Iowa, a group of unhappy Collins engineers called, and in June, we would open our second design center in Cedar Rapids. Rockwell would threaten to sue us over that. Later in the year, we would open a third design center in Billerica, Massachusetts, near Boston, that would employ 35, and we had a fourth planned for Chandler, Arizona, near Phoenix.

In January, we had learned that TRW was negotiating to buy a British firm, LucasVarity, which had been created by the merger of a U.S. auto parts manufacturer and a British aerospace company. We had no idea then that this potential buyout might affect us, but it would.

The deal was agreed to by the end of March, although it would take two more months for the sale to be completed. TRW was paying $6.8 billion in cash for LucasVarity. That would leave the company with huge debt and an urgent need for money. And we were notified that one way they planned to raise that money was by selling our stock.

Although TRW had sold 250,000 shares at our secondary offering in January (which should have given us a hint of what was to come), it still owned nearly 11 million shares after the stock split on March 17, about a third of all outstanding shares.

If it started selling huge blocks of stock on the open market, it wouldn't escape the attention of institutional investors and market analysts who might begin asking, "Does TRW know something that we don't?" It could spark a selloff that would send the stock price into a tailspin. Even if that didn't happen, it could push the stock price down, or keep it from growing.

We also were concerned that if word got around that such a huge amount of stock was available, it might make us vulnerable to a takeover.

Despite our concerns and our hope that the stock might be placed privately with institutional investors, TRW moved forward with a plan to put the stock on the open market.

After the price of our stock had been cut in half by the split, it drifted downward, closing at $45.75 on April 1. Two weeks later, analysts began predicting that we would have a strong fourth quarter, in part because our fab was increasing production and margins. One praised our "conservative, methodical management." *Motley Fool*, the Internet investor magazine, proclaimed our stock "one of the hottest to own over the next six months," and the price shot upward.

After we announced our fourth-quarter results on April 20, the stock leaped 27 percent the next day, closing at $63.63. Revenue for the quarter was $56.5 million, up 374 percent from the same quarter of the previous year. Profit was $9.94 million. Nokia had accounted for only 57 percent of sales. For fiscal '99, sales were $152.9 million, with a profit of $19.6 million.

Stock prices were soaring at this point, and ours continued to rise until TRW sold 440,000 shares from April 23 to 28. Over the following month, they sold another 940,000 shares, sending the price plummeting. On May 25, it closed at $39.88, and the following day, when I spoke to the Greensboro Chamber of Commerce, everybody wanted to know what was happening to the stock.

"I can tell you the company is on a solid footing," I assured them. "We're not even beginning to run out of growth."

At the beginning of June, over eight days, TRW sold another 1,554,000 shares. On June 7, we announced that we had begun production in Fab Two three months before the projected date, and the stock started climbing again. It had been at $40 five days earlier, but it closed on this Monday at $55.50.

We now owned not only the world's largest gallium arsenide HBT fab, but the world's largest molecular beam epitaxy fab. And we were preparing to announce that we were about to give the world an even larger gallium arsenide HBT fab.

In a lengthy article on Friday, July 2, *News & Record* reporters Ben Feller and Peter Krouse unveiled our plans to build Fab Three. It appeared under a big, bold headline across the entire width of page one: "RF Micro could add 800 jobs."

"The project would be one of the biggest economic developments in recent county history, given its dollar value and potential for cherished high-skill jobs," the story noted.

There was a 50-50 chance that the fab would be built near our other facilities, Bill had told the reporters, but other areas were under consideration.

"Whatever we do, it has got to be fast," he said, "because we just don't have time to mess around."

We actually were considering a site in Dallas, near a Nokia plant, as well as potential sites in Scotland and Ireland, where government incentives were especially appealing.

Nokia didn't want the fab in Greensboro. It was anxious about having all production in one spot where a calamity such as a tornado, earthquake, or—especially since we were in the flight path of the nearby airport—a major airliner crash, could suddenly halt its supply of vital parts.

Still, we wanted to put it on what was becoming our campus at Deep River Business Park, because it would be quicker and easier to build, as well as handier to manage. But that would depend on the tax breaks and other incentives we could get from the city, county, and state. This time, because time was a pressing priority, we had set a deadline of July 20 for their responses.

I was just thankful that this time I wouldn't be involved in those testy political negotiations. We had hired a consultant, Patric Zimmer, to handle that job.

On Monday, July 19, the day before we were to issue first-quarter results, our stock had been above $80, but it took a sharp dive on Tuesday, losing $11.25. Most technology stocks dropped due in part to profit taking, but our loss also could be attributed to worries by some analysts that our quarterly report would not be up to expectations. They were wrong.

Our sales for the period were $62 million, nearly triple those for the same quarter a year earlier. We had profit of $10.4 million. We issued the report after the close of the market along with another bit of news that would make headlines. We would split our stock for the second time in five months on August 18.

These announcements came on the same day as our deadline for state, city, and county governments to decide on the tax breaks and incentives for our new plant, and that night the Greensboro City Council approved $2.5 million. The state, and this time even the county, already had come through.

Two days later we announced that management would recommend to the board that the new fab be built in Greensboro, although we couldn't predict the board's response. That decision would be made following the annual shareholders meeting on Tuesday, July 27.

In the July 28 issue of the *News & Record*, Peter Krouse wrote that our stockholders meeting seemed like church, a joyous one.

"But it wasn't spiritual hope filling their hearts this day," he wrote of the congregation, "simply the delight that comes with an earthly investment that pays big returns."

Our new fab had been the most important factor in our spectacular growth, Dave told the investors, going on to assure them that the growth had just begun.

"We're not going to be a bit player," he said. "We're going to be one of the big guys."

After the shareholders meeting, the board gathered to decide where the new fab would be built. We had no concerns about how the vote would go. We already had announced a press conference for the following day to which many dignitaries had been invited.

Dave loved to joke around, and he couldn't resist on this occasion.

"I've got good news and bad news," he announced.

The good news was that the board had approved building the fab. The bad news?

"It's going to be in Pittsburg."

No laughter. Only silence, except for a few gasps from politicians, governmental and chamber of commerce officials.

"Just kidding," Dave added quickly, provoking more sighs of relief than laughter.

The new fab would be built directly behind our first fab, and construction would begin on Monday. He expected actual costs of the plant to be closer to $250 million than the $200 million previously announced.

Later, in response to a reporter's questions, Dave summed up what we all thought about the incredible growth we were experiencing.

"We feel that we are causing change in this industry, not reacting to change. Yet we are still amazed at the growing demand for our products. We had faith that we would make it big, but this big? We've been blessed."

Early in August, we hit a memorable milestone, shipping our 100 millionth power amplifier.

That month we also leased 77,000 square feet of a 102,400-square-foot building adjoining the site that we now were preparing for Fab Three. By the end of the month, we had begun a $10 million revamping of the building to turn it into an automated packaging plant. Here we would be able not only to package our prototype chips but also modules—several chips and other components encased in a single plastic-covered circuit ready to be plugged into a phone, or other device. Modules, we felt certain, would become a major part of our future business.

We actually would have the capacity to do some production work in this plant, which would help us keep a competitive edge with the packagers in Asia who now prepared most of our chips to fit into circuits. We hoped to have this operation up and running in November.

When that happened, we would for the first time be able to make a chip from raw materials, package it, and ship it to a customer without the assistance of any other company.

August also brought the culmination of my months-long campaign to get us the technology that IBM was denying us: silicon germanium.

Only a handful of companies had this technology. Among them were ST Microelectronics in Grenoble, France, one of IBM's major competitors; Infenion in Munich, Germany; and a couple of smaller companies in Canada.

Over recent months, I had gone to all of these companies and negotiated tentative partnership agreements that would grant us silicon germanium.

I had all of those in hand by August and went back to IBM, whose technology we preferred, because it was technically superior. I called the top executive of IBM's microelectronics division, John Gleason, but he was at a trade show in Las Vegas. I reached him there, and he invited me to come on out and meet with him.

We had an amiable session. I told him how much we treasured our long and fruitful relationship with IBM and how much we wanted to continue it. But IBM's decision to deny us silicon germanium had forced us to seek other alternatives, I said.

I told him about the companies that were willing to accommodate us and explained that the agreements we had reached with them were conditioned on terminating our relationship with IBM and granting all of our silicon business to the company we chose. We didn't want to do that, of course, but we would have no other choice unless IBM reconsidered.

John said that he understood, thanked me for granting IBM a second opportunity, and I flew back to Greensboro. A few days later he called to say that IBM would grant us silicon germanium, and I began working with Ron Soicher and Michael ConCannon at IBM to hammer out the terms of the deal. In the end, we got silicon germanium for a cash payment of $1 million.

I signed the contract on September 9, but I wanted to go back to all of the companies with which I'd previously negotiated to let them know that we had chosen IBM, and we wouldn't announce the agreement until October 12. Even before we signed the agreement, we already had our first silicon germanium part ready for prototype production for the cell phone maker Ericsson.

We have since sold millions of parts in silicon germanium, all made by IBM, and we are developing many more. Our continuing partnership brought hundreds of millions of dollars to IBM. Sometimes companies just have to be nudged a little to realize what's in their own best interests.

A rapidly growing force became a huge part of the economy in 1999. The Internet, it was said, was going to dominate our lives, and Internet companies were jumping up like popcorn in a hot kettle. The Internet, along with quickly expanding cable TV networks, and the predicted coming of the new, Internet-connecting, data-devouring generation of cell phones, all had come together to make fiber-optics a hot commodity.

Fiber-optic cables contain thousands of hair-thin strands of fiber from which light cannot escape. Using transmitters and receivers, light signals can pass through these strands, carrying far more information or conversations, far faster, with less interference and less required amplification than standard electrical lines such as copper.

Fiber-optic cables were being laid all over the country at an astonishing pace, and companies dealing with fiber-optics were growing as fast, or faster, than ours. This was a market in which we saw potential for gallium arsenide HBT chips. The transmitters and receivers on fiber-optic cables could use them, as could the cable TV set-top boxes, signal boosters on cable lines, and other devices

But our license with TRW allowed us to use gallium arsenide HBT only for wireless products. We would need to expand the license to make parts for this new market. We also wanted to be able to make wireless parts for the coming broadband technology that used higher frequencies than our agreement allowed.

In May, two of our board members, Al Paladino and Erik van der Kaay, who had negotiated the first license, met with TRW officials to feel them out about this possibility. But months passed with nothing coming from it, and we began to believe that TRW was holding out for perhaps a bigger and more lucrative prospect.

In mid-summer, Bill, Dave, and I decided that we would have to start pressing for a response so that we could determine whether or not this was a feasible proposition, and I agreed to take on the project.

Bill wasn't optimistic about our chances. He believed it would be all but impossible to wangle a deal out of TRW at an acceptable price, considering what we had to go through to get the first license.

I jokingly offered to wager that I could do it, and Bill accepted.

"What are you willing to put up?" I asked.

"What do you want?"

"I need a new bulldozer."

"All right," he said, "if you get this license, I'll buy you a new bulldozer of your choice."

"Uh..." I hesitantly asked, "what do I have to give you if I don't come through?"

"How about a steak dinner for me and my family?"

I quickly did the math. Steak dinner—bulldozer. Seemed a little weighted in my favor.

"That sounds reasonable," I said.

I began by calling my old friend Bob Van Buskirk, who had left TRW. I wheedled him for any intelligence that might be helpful, especially about the people with whom I'd be dealing. Then I began the arduous process of pushing a slow-moving, government-endowed corporate bureaucracy to action, a process not unakin to getting that final, huge, pointed block of stone to the top of one of the great pyramids in days of antiquity.

I was dealing primarily with Wes Bush, vice president and general manager of the telecommunications division at TRW's Space Park in Redondo Beach, but I wasn't getting far.

"That's an interesting proposal," he'd tell me. "I'd be willing to study that. Let me get back to you."

A week would pass, then two, and I'd call again and get a similar response.

Beginning on September 23, TRW again began selling our stock, moving 1.73 million shares over the next eight days, pushing the market price steadily downward. On October 6, TRW sold another 3.2 million shares.

We were distressed about this development, particularly our board members, and we decided that we had to do something to try to control the situation and prompt action, one way or the other, on the proposed license revision.

TRW agreed to a meeting at our offices on October 25. TRW's team consisted of Wes Bush and three executives from corporate headquarters: Chief Operating Officer Ron Sugar, Treasurer Ron Vargo, and Chief Financial Officer Karl Miller.

Karl Miller delivered an assessment of TRW's financial situation and the necessity of selling off assets, which included the bulk of its investment in RFMD. Ron Vargo talked about how to get the most money from the stock TRW still held, which included the possibility of private placement by investment bankers, with whom it was in discussions.

Wes Bush spoke about a new strategic alliance between the two companies that would involve some expansion of our license agreement.

Dave and I presented our position, which essentially came down to this: Private placement of the stock would allow the market price to stabilize and rise on its merits. Considering the immense success of our alliance so far, a new technology agreement would benefit both companies.

No conclusions came from this meeting, and we began pulling together a detailed presentation on what took place for our board's regular meeting two days later, where we would decide the steps to take next. I also was preparing a little surprise for our board members.

The two-story addition to our headquarters building, into which we had moved late in the winter, had two new conference rooms, in addition to the conference room in the original unit. The new fab also had a conference room. This caused confusion, because employees often would misunderstand which room they were supposed to go to for a meeting. They would get to one room only to discover that the meeting was in another, sometimes resulting in late arrivals.

The way to solve this problem, I decided, was to name the conference rooms. And who better to name them for than the outside members of our board of directors? It would be a nice way to honor them. And if we honored them, it might make them less likely to give us a hard time.

Our board at this point consisted of only six members: Bill and Dave; our three original investors, Al Paladino, Walter Wilkinson, Erik van der Kaay; plus Terri Zinkiewicz, finance director of TRW Space Park, who had succeeded Tim Hanneman in the TRW seat.

I always liked to provide as much entertainment as I could get away with for board meetings, and for the one scheduled on October 27, I planned an elaborate naming ceremony.

The rooms, I decided, should be named completely at random, and the best and showiest way of doing that would be to have a drawing.

I got two fish bowls that we used at trade shows for collecting business cards and offering treats and polished them with window cleaner. Kathy typed the locations of the conference rooms on one sheet of paper, the names of honorees on another. She cut out the names and locations in strips of identical sizes and folded all in exactly the same way. One set of strips went into one fish bowl, the other into the second. We wanted to make certain that everybody could clearly see that this was on the up-and-up.

I know that if I said here that it never occurred to me that it would be not only high irony but really funny if the conference room in Fab One ended up named for Al, some people might find it hard to believe.

Nothing I can do about that.

I should add, however, that although Al had opposed building Fab One, he fully supported Fab Two, which we now were operating, and Fab Three, which we were building.

When my time came at the board meeting, I announced my purpose, brought out the fish bowls, and proceeded with the drawing. I already had ordered brass nameplates to place on the walls outside the rooms.

I drew first from the location bowl, and I turned my head away so that I couldn't see the folded slip for which I was groping. I announced the location of the room, then I drew and announced the person for whom it would be named.

I can't remember now exactly how this unfolded, but I believe that the first two rooms named were in the newly opened addition to the headquarters. That left two to go, one of them in the fab—and Al's name was yet to be drawn. This presented just what I wanted: a moment of high drama, not unlike that point in the Miss America pageant when only two contestants are left clinging to one another waiting for glory—or despair.

I drew from the location bowl and unfolded the strip of paper.

"The fab," I said, and showed the slip as proof.

Then turning my head away, I fished in the other bowl. Out came the slip of paper. I opened it with agonizing slowness, saw the name, and did a doubletake. I couldn't suppress a grin.

"Albert Paladino," I announced with all the dramatic flair that I could muster, and the room erupted in laughter.

Never before had I gotten such a big laugh at a board meeting, and I haven't gotten a bigger one since. Thankfully, Al, in all his glory, was laughing, too.

"This is your way of punishing me, isn't it?" he said. "You've never pulled a more clever trick. How did you do it?"

"I didn't cheat," I vowed. "Everybody saw it."

Even today, Al still brings it up.

"I know you set me up," he says. "I just don't know how you did it."

"Hey," I keep telling him, "you saw it with your own eyes."

On Friday, October 29, at the board's instruction, I faxed a letter to Wes Bush. I began by stating what a hugely successful partnership we had created.

"Unfortunately, TRW's decision to explore the sale of a significant amount of RFMD stock, coupled with TRW's recent stock sales and our inability to come to a quick agreement on HBT license expansion, has put considerable strains on our relationship."

Our board, I went on, had decided that "the current situation is not acceptable and that prompt action needs to be taken...to preserve our relationship and avoid continuing adverse impacts to RFMD and its shareholders."

Our own expanding capacity soon would leave us unneedful of TRW's foundry, which was an issue for the company, but the board had authorized me to allow production to continue there for three years at a minimum rate of 5,000 wafers annually as part of an accord for a license expansion.

We wanted to make the private placement of TRW's remaining stock part of a new strategic alliance agreement, and we were willing to do whatever we could to help TRW bring it about, including sending our senior management on a road show to make presentations to investors.

I stressed that we needed to move fast before the market reached its own conclusions about our relationship and did further damage to the stock value. I asked for a response by Monday.

That letter broke the logjam. Wes responded with solid terms for the license expansion, but spoke only vaguely of cooperation in coordinating the stock sale. In essence, TRW would grant us a license only for broadband TV products in exchange for continuing production in their foundry and warrants for another million shares of our stock, which if the price could be stabilized and continued to grow as it had in the past year, could be immensely valuable.

At least, we had a place now to start negotiating, and over the next 10 days offers and counteroffers flowed back and forth. Finally, on November 12, Wes came to Greensboro with lawyer in tow, and the two of them faced down me; Suzanne Rudy, our treasurer; and Jeff Howland, a member of the law firm that represented us, in a long day of hard negotiations.

Wes was being so tough and so firm that at one point I despaired that this was an exercise in futility. I'm slow to anger, and it rarely happens, but this time it did, and it startled Suzanne and Jeff. I suggested that we take a break, but I figured that this was the end.

When we all came back after the break, however, the tone had changed. A little movement began, and a few hours later we had an agreement. We got our expanded license. TRW got warrants for 250,000 shares of stock at a cost of 75 percent of the average price of the stock over the last 10 days in December. It would receive warrants for another 500,000 shares conditioned upon the sale of a certain amount of the products (these never would be redeemed), plus our continuing purchases of minimum annual numbers of wafers from its foundry.

Our big disappointment, and continuing worry, was that TRW would not commit to a private placement of the stock. It agreed only to maintain a 10 percent equity in the company until May 1, 2001.

Two days later, the contracts were signed, and on Monday, November 15, we jointly announced our new alliance. We also announced our new division for broadband products. It now had 10 employees, but we would be adding another 18. None of us could have foreseen that within a couple of years the fiber-optic market would collapse and that this division would be disbanded.

On Friday, November 5, our stock had jumped $7.38 to close at $62.25 after an article in *Barron's*, the so-called business bible, declared that companies making components for wireless communications were good buys. That day we also had released news that once again we were expanding our facilities. We had leased a new, 120,000-square-foot building in the business park on Gallimore Dairy Road, where Fab Two now was in operation, that we intended to turn into a new testing lab, an absolute necessity if we were to continue our ever-increasing production. We planned to complete only half of it in the beginning, and to have it in operation by June.

After our announcement of the new pact with TRW, the stock rose almost $5 to close near $78, an all-time high, considering the two stock splits. But three days later, on November 18, TRW began selling stock again, bringing our price rise to a screeching halt. Before the year was out, TRW would unload another 4 million shares, and although the stock price sank, it held at a higher level than we anticipated.

Not long after the agreement with TRW was completed, I strolled into Bill's office.

"Looks like we're going to be doing some shopping," I told him.

"I guess you're right," he replied. "Have you been looking around?"

"As a matter of fact, I have. Come on and I'll show you what I found."

I drove him up the road to Bestracs, where I pointed out a beautiful new yellow John Deere 450H bulldozer, the top of the line at the time.

Bill didn't say much. He just took out his checkbook.

To rub in my victory a little, I decided that we needed to have a public key-passing ceremony. On the day that Bestracs was to deliver the bulldozer to my farm in Randolph County, I got the store to make a detour by the company. This magnificent piece of machinery was chained to a Lowboy trailer, and the truck driver parked it by the curb right in front of the company entrance.

I'd had a big sign made that said "Billdozer," and I taped it across the blade. Then we made an announcement on the company public address system that everybody was invited to see Bill ante up, and people came pouring out of the building.

Christmas was getting near, and I was wearing a Santa Claus hat with my jeans, work shirt, and brogans, even though Bill was the actual Santa here.

Bill climbed onto the trailer beside the bulldozer, made a little concession speech, and held the keys up for me to come and get. I stepped up with great enthusiasm for my prize, but a board gave way, my foot suddenly dropped 18 inches to the pavement, and I sprawled head-first onto the trailer bed.

I always like to entertain, but I hadn't prepared for pratfalls. I understand the importance of entrances, mind you, but as spectacular as this one was, it lacked somewhat in grace. Several people realized this wasn't part of the act and rushed to assist me as I attempted to crawl out of the hole.

"Are you okay?" they kept asking.

"I'm fine," I insisted, grasping desperately for my dignity. "No, really, I'm fine."

Actually, I wasn't. I was in intense pain, but I wouldn't have acknowledged it even under threat of being run over by a bulldozer.

I claimed the keys and made the presentation I had prepared. One of the reasons I had wanted to do this was so that I could publicly recognize Suzanne Rudy and Jeff Howland, who had worked so hard to bring about the TRW deal. I presented them with gifts and thanked them for their work.

Nobody was more thankful than I, though, when this event ended. My leg was throbbing, and I hobbled back to my office, pulled up my jeans, and found all the skin missing from my shin, the blood already clotting. Kathy fetched the first-aid kit, I doused my wounds with searing antiseptics, wrapped the bottom half of my leg in bandages, and hurried home to take my bulldozer for a spin.

To this day, Bill still denies having anything to do with that loose board.

At the beginning of 1999, our stock had a market value of $792 million. At the close of the year, that amount had climbed to $5.4 billion, a figure almost impossible to believe. We only can guess what it might have been if TRW hadn't been selling stock throughout the year. This year had taken us on a wild ride, indeed, but as events would turn out, we really hadn't seen anything yet.

11
Spiraling Insanity

As the final minutes of the unbelievable year of 1999 ticked away, millions of people were anxiously anticipating the predicted disaster that "Y2K" was to bring to our technologically interconnected world.

People huddled with stores of batteries, canned foods, bottled gas and water, awaiting the chaos that would ensue when the dawning of the new millennium rendered computers useless and civilization unable to function.

As vivid a reminder as the Y2K scare was of how dependent we had become on information technology in such a short period, that fear proved groundless. The year 2000 glided in with only a few glitches, leaving people to deal with their stocks of panic supplies, blissfully oblivious to the true disaster that the year held in store for stocks of a different kind.

Only in hindsight do the events early in that year appear to be as absurd as they actually were. Now, it's easy to see that the financial markets were wildly out of kilter, and a reckoning was certain and soon. But then it seemed wonderfully believable, nothing more than due good fortune, especially to us.

People far more savvy about economics than I—for one, Alan Greenspan, chairman of the Federal Reserve Bank—have said that it isn't possible to recognize a market bubble as it drifts ever higher. Only when it bursts does it identify itself, and even then not immediately, but grudgingly over time.

As usual, a few seers were pointing at this bubble and screaming warnings, but in the exuberance and euphoria of the time, few were listening. Everybody was making too much money, or so they thought in all too many cases.

Most of the so-called experts—the stock analysts, the economists, the business news reporters—were joyously telling us that we were at one of history's great turning points. Just as the Agricultural Age had given way to the Industrial Age, the Industrial Age now was shifting into the Technological Age. We just had the astounding good fortune of being in the right place at the right time.

You couldn't pick up the business page of a newspaper or flip the channel to CNBC without being informed that we had created the "new economy," an economy with totally different "paradigms," whatever they may be, an economy in which the old rules no longer applied. We finally had erected a golden stairway to riches that anybody with a few dollars to invest could climb.

We had been in a bull market for so long that supposedly serious people were predicting that it wouldn't end, that we had created a "recession-proof economy" in which the Dow Jones Industrial Average soon would top 15,000, then sail on to 30,000 and beyond.

And the Nasdaq? Well, that was where most of the companies that had inspired this revolution were nestled: Microsoft, Intel, Cisco, Dell, MCI Worldcom, to name the big ones. It was where most of the technology stocks were traded, including ours. Nasdaq was where the action was, the center of the economic universe.

To get a picture of the craziness that actually was going on, you need only look at what was happening on the Nasdaq market in 1999. From October of 1998 through 1999, it grew 130 percent, the greatest spurt of growth of any stock exchange in history. At the end of 1999, the Nasdaq stocks alone were worth more than all the stocks on the Nikkei Exchange in Tokyo at the end of 1989 at the peak of the biggest market bubble in history to that point.

Nasdaq was the market of choice for new companies launching initial public stock offerings, and venture capitalists were taking advantage of it as never before. Prior to 1999, VC firms on average were funding about 1 percent of the companies whose business plans they considered. In 1999, that number increased more than tenfold.

Most of these new companies were Internet companies, what came to be called the dot.coms. The Internet promised to change all of our lives forever, deliver us to that great new paradigm in the sweet by-and-by, and these were the companies that were to take us there. Scores of dot.coms were popping up almost daily, and VCs were falling over one another in the scramble to lard money on them.

Often these companies consisted of little more than a premise, with no customers and no real hope for profit, but VCs were funding them on the spot, then rushing out the IPOs while they were hot. In 1999, IPOs on Nasdaq quadrupled from the previous year, their value more than $50 billion. On average, these stocks were increasing 60 percent on the day of issue, but some were doubling, and more. Almost any day you could read or hear about new instant Internet millionaires.

Investors were grabbing up stocks of every company that had anything to do with the Internet, and at whatever price. So frantic did the pace become that in December more than two billion shares were being traded daily on the Nasdaq, and that would continue and grow into the new year.

At year's end, the average Nasdaq stock had tripled in 1999. Ours didn't quite match that, starting the year at $27 after split adjustments, and closing at $69. Our former customer Qualcomm fared considerably better. Its stock began the year at $26 and grew twentyfold by the end.

The stock markets continued their dizzying rise in January. Our stock jumped $10 on January 3, but that came to a screeching halt the following day, because TRW started selling again. Over the next month, it sold nearly four million shares, flattening our stock price, while the stocks of other technology companies kept shooting skyward.

Despite that, in January *Bloomberg Personal Finance* named us number nine on its list of the 100 hottest stocks, although many others were doing far better than ours at that point.

On January 14, the Dow hit a record high—11,772—then slipped into a decline that would continue for two months. But nothing appeared to be able to stop the Nasdaq. Powered by more and more dot.com IPOs, it was a whirlwind with tornadic certainty.

In this market, the unbelievable quickly became reality, and a prime example came at the end of January, when America On Line, the Internet access company, consumed the media giant Time Warner. AOL had less than a fifth of Time Warner's revenues but three times its value.

We were so busy with our own explosive real growth and the problems it brought that we had little time or attention to pay to the madness around us. Still, we couldn't help but notice what was happening on the Nasdaq, especially with the Internet companies.

I remember talking and laughing about it with Bill, Powell, Dave, and Dean. The whole thing was ludicrous to us. We had to work hard for years to get to our IPO, only to see the price rise slightly before falling back. But here were these instant companies going public with little more than an idea and only the faintest of effort and the shares took off like rockets.

We were a real company, with real customers, real products, rapidly growing sales and profits, and a market that seemed without end, yet the stocks of these fantasy companies often were valued far higher than ours. Amazon.com, which went public two weeks before we did, and would eventually become one of the rare Internet success stories, was valued at more than $30 billion in 1999, and its founder, Jeff Bezos, even made the cover of *Time* magazine as newsmaker of the year, although the company had $400 million in losses and was still years from even a hint of profit.

We never dreamed that these dot.com companies possibly could affect us. If we had, we wouldn't have been laughing.

At the beginning of February, we were the subject of a lengthy, flattering article in *Electronic Business* magazine called "The Little Powerhouse" by Paul Gibson. It began this way: "By rights, RF Micro Devices Inc. should never have emerged as an independent business, let alone succeeded so well."

Gibson went on to state: "RF Micro's success could be the stuff of a business school case study. It not only explains how a nimble and determined start-up can run circles around a handful of big-name competitors, including the likes of Motorola...Phillips Electronics...and Hewlett-Packard...but it also illustrates how Analog Devices...and TRW...missed a golden opportunity to strike it rich because they were too timid."

These were not things that we would say ourselves, of course, but there was a certain satisfaction in reading them.

At our board meeting on February 1, I made a presentation in which I took great pride. After all these years, we were regaining Qualcomm as a customer.

I had kept up with Don Schrock at Qualcomm. We talked periodically by phone. By this point, Qualcomm had gotten out of the phone-making business, but still was manufacturing chipsets for CDMA phones and selling them to licensed companies. It was making the power-amp modules in silicon for these chipsets but wasn't getting the performance out of them that it wanted.

Don and I were talking about this one day, and I suggested that we should hook up again and solve this problem. It was clear now that with our new fabs we would have the capacity to supply whatever quantities Qualcomm needed. Don agreed to meet me in San Diego to talk about it, and we later got our engineering teams together to work out the technical problems.

We would design a power amp module to fit into Qualcomm's chipset and make it easy for its customers. Qualcomm would get a technically superior power amp, and we agreed to price it so that we would be splitting our normal profits with Qualcomm. The huge numbers of chips we would be selling would make that feasible. We set up a special support group of 16 people to deal with the project.

We all thought that getting Qualcomm back would be a great boon to the company, and it would prove to be so.

Our stock was at $80 on the day I told the board about the new deal with Qualcomm. Tech stocks were showing rapid gains at the time, and after TRW stopped selling our shares on February 4, our stock began rising along with those of other tech companies. On February 15, it hit $100 before closing at $97. It remained close to the $100 mark until Friday, February 25, when it took its biggest jump to that point.

Two securities firms, First Union and PaineWebber, initiated coverage of our company that week with "buy" recommendations. After PaineWebber's announcement on Friday, nearly seven million shares of our stock changed hands, and the price shot up to $120 before closing at $115, a jump of almost $14 in a single day.

I was at a trade show in New Orleans the following Monday, February 28, and it was there that I announced our new alliance with Qualcomm. Analysts hailed the news.

"They were put in a penalty box," Samuel May of U.S. Bancorp Piper Jaffray said of us, "and once you get in, it takes a lot of work to get out."

That day the stock set another record, rising to $149 before closing at $139. More than 10 million shares were traded. On Wednesday, March 1, the price closed at $149, and we were on a tear.

Six million shares were traded on Thursday, and although the price rose to $164 at one point, it closed only a dollar higher. But Friday proved to be another record day, with the price closing at $175.

The new week brought more stunning advances. The price rose briefly to $194.50 on Monday, the highest point to date. At that moment RF Micro Devices was valued at $15,793,456,000. If a person had bought 100 shares of our stock when it was at $10 a couple of years earlier, that $1,000 investment would have been worth $738,000 at that point.

Less than nine years earlier, Bill, Powell, and I hadn't been able to entice even $10 out of any investor, and now the company we created was worth nearly $16 billion, a figure that defied an adjective adequate to describe it.

Was that a reasonable value? Not if you considered that in little more than two months the company hadn't substantially changed while its market value nearly tripled.

Should that have told us something? If nothing else, it was a sterling example of what an extraordinary time our economy was passing through.

On March 10, the Nasdaq hit its highest point ever—5,048.62. It had doubled since the previous summer. From that day, it began drifting downward, as our stock had already.

But as if to demonstrate what a wildly unpredictable period this was, a week later, the sagging Dow leapt 800 points over two days, yet another record.

• • •

The Los Angeles Times offered a striking illustration of the presumed irrationality of this period in its business pages on Sunday, March 26. The article by James Flanigan was about TRW, but it began with a comparison of TRW and RF Micro Devices.

Our company, Flanigan pointed out, had roughly $300 million in annual sales (it actually was $11 million less than that), and our stock had by then fallen to about $160, which was 290 times our earnings per share.

TRW, on the other hand, had $17 billion in annual sales, but its stock was selling at only $56, or 10 times its earnings per share.

Yet our market value far exceeded TRW's.

Incredibly, it was twice as much as that of our technology partner and major investor, the huge, long-established defense company to which we had turned to get our start.

But that wasn't the point of this piece.

"Is this another story of a 'new economy' small company using high technology to leave an 'old-economy' warrior in the dust?" Flanigan asked.

"No, far from it. It's a story of change and adaptation beyond the current clichés that tell how industries and technologies arise and how smart people find a way."

Flanigan's focus was TRW's acumen in commercializing the technologies it had developed for NASA and the military and intelligence agencies, and its partnerships and investments with us and half a dozen other new technology companies that offered bright prospects for TRW.

What the article failed to recognize was that until I made that desperate call to TRW in the fall of 1991, seeking a way to produce that first failed power amp for Nippondenso, TRW had shown little success in commercializing its technologies. It was Bill's vision and our success that had led TRW down this lucrative path.

So it could be asked, I suppose, whether it really was so strange that in this market—which valued vision and potential growth over assets, profits, experience, and all else—we were considered to have far greater value than our patron.

The first real rip in the bubble in which this astounding "new economy" was floating came on Monday, April 3, when U.S. District Judge Thomas Penfield Jackson in Washington found that Microsoft had violated anti-trust laws in attempting to capture the market for software that allowed Internet access. Microsoft's stock plunged drastically, taking other technology stocks with it. Our stock closed that day at $116, down nearly $80 from its high mark less than a month earlier. It dropped daily after that.

But the big hit didn't come until 11 days later. April 14 would be declared "black Friday" in news reports. On that day, both the Dow and the Nasdaq suffered record one-day losses—616 points for the Dow, 356 for Nasdaq—a collapse akin to the October 1929 crash that touched off the Great Depression.

"Nightmare on Wall Street," headlined TheStreet.com.

"The trading week died at 3 p.m. EDT and fell into its grave face first," wrote reporter Eric Gillin.

TheStreet's reports were filled with phrases such as "ruthless sell-off," "devastation," and "no relief in sight."

Perhaps Anthony Perkins, chairman of Red Herring Communications and co-author of *The Internet Bubble: Inside the overvalued world of high-tech stocks—and what you need to know to avoid the coming shakeout*, best summed up the effects of the day in *Salon*, ironically an Internet magazine that soon would be facing its own problems as a result of this day's crash.

"Clearly, the bubble has burst," said Perkins. "I think things will never be the same. I think a lot of the Internet companies that shouldn't have gone public have been found out, and you won't see them go back up. What's happened is a lot of people have lost a lot of money. I think it's going to scar people for a long time, and they're going to realize that it's now back to fundamentals. It's back to rationality."

I doubt that on this day anybody realized the true extent of what had happened. The dot.com bubble, as this one would become known, was in many ways like the gold rushes that had followed other technological developments—the telephone in the nineteenth century; Marconi's introduction of the wireless telegraph shortly after the turn of the twentieth century (Remember *Success* magazine's alarming series of articles in 1907—"Fools and Their Money/The Wireless Telegraph Bubble?"); and commercial radio in the 1920s. All of these had produced stock market bubbles that had burst with drastic consequences.

The dot.com collapse was different from these only in scope. Far more people and many more companies would be hurt by this calamity. Vastly greater investment sums would disappear, along with hundreds of companies and thousands upon thousands of jobs. The dot.com burst would be historic, the greatest loss in stock value of all time.

But we had no way of knowing that on that dark Friday, April 14. Our stock closed that day at about $79, down a little more than $18, just $10 more than it had been at the beginning of the year before it vaulted to such giddying and unrealistic heights. We looked at this as a correction, a return to sanity. The market simply was reacting to the falseness of the time, an attempt to shake off the companies that had no business being there in the first place.

If the dot.coms started falling away, our response would be good riddance. In our minds, they simply were creating problems for genuine, productive companies. It didn't seem possible that their disappearance could have anything to do with us. We had more business than we could handle, and were adding people and facilities as fast as we could. We had many new products in development, big orders for months on end, and strong predictions of ever-growing business. We couldn't conceive that the disappearance of these ephemeral companies could harm us.

Our belief that this was just a correction was reinforced on Monday, April 17, when the market showed signs of recovery. Our stock surged back to $94. The following day we released our fourth-quarter report. We had revenue of $84.8 million, earnings of $14.6 million. Revenue for fiscal 2000 reached $289 million, nearly double the previous year. Profit was two and a half times the previous year's at $50.1 million.

Clearly, we were a company on the move, with increasing profit and remarkable growth. But our stock dropped $13 on the day after this announcement.

We had finished our new headquarters building at the beginning of the year, and had moved into it at the end of January. No longer did we have to worry about what potential customers might think when they came to visit.

We had a large, beautiful, well-appointed reception area with lots of greenery, a rotunda with a glass dome, spiral stairways, display cases for awards and memorabilia, a cafeteria big enough to seat several hundred, an impressive board room with a huge video screen, and all the latest equipment for teleconferencing.

For the first time, I had an office with room enough for something other than my desk and a chair or two. It was in a corner on the second floor with wide views in two directions. But it did have one slight drawback. It was directly under the pathway of jets taking off from the nearby airport. I had my desk placed so that my back would be to the windows facing the runway. If one of those jets was going to hurtle through them, I didn't want to see it.

We had erected this structure ourselves for efficiency's sake, but from the beginning we had intended to sell it to a development company and lease it back. We put it on the market in April and quickly got dozens of offers. But reporters got wind of it, and associating it with the recent stock collapse, they began to wonder if the sale was a sign that the company was in trouble.

Dave assured the first reporter who called that we were selling the building because we could get a better return on the money by investing it in new technology. But that didn't stop the questions.

"The value of a company comes in what's inside the building, and not in the building itself," I told another reporter who called, Richard Craver of the *High Point Enterprise*. "It made sense for us at the time to build the building ourselves. It now makes sense to sell and lease it back...and it keeps us out of the real estate business as well."

At this point, we had 871 employees and were adding as many as 20 a week. Every day now we ran into people in the hallways we'd never seen before. Such big and rapid growth causes problems for any company, but we hoped that all of these new people could be absorbed without any threat to the values we had established, and we had taken steps to assure that.

For years we resisted hiring a human resources manager. (Forgive me, but I still lapse now and then and think of such a person as a personnel director.) In the early years, Bill, Powell, Dave, and I, and sometimes Dean, interviewed most of the people we hired, but that became increasingly difficult. We often interviewed half a dozen people for a position, and the logistics—getting them airline tickets, rooms, and the like—became a nightmare, frustrating our administrative assistants.

We were reluctant to create a human resources department because of our experience at Analog Devices, where that department was highly politicized. We didn't want the consequences that produced. When it became obvious that we had to create such a department, we went looking for somebody who would make certain that never happened.

That person turned out to be Elizabeth Moore. We found her at ITT in Roanoke in the fall of 1996, and luring her away so angered some folks at ITT that the company dropped us as a customer. (Later, we also hired her husband, Forrest, to head our information technology department.) Liz, who retired late in 2003, agreed with us that a human resources department should have no agenda other than to implement the policies of the executive staff, serve the employees, and give them voice. Liz did a beautiful job setting up the department, hiring new people, establishing standards, and making everything work smoothly.

We were growing so fast that people who had been with us for only a relatively short time quickly became senior staff. From the beginning, we had concentrated on hiring high-energy people who shared our philosophy of the business. And, not to pat ourselves on the back, we did remarkably well.

In our earliest years, we hired many of the people who would become company leaders. In 1992, we got Kellie Chong, who would become director of corporate engineering, and Bob Hicks, director of testing operations. In 1993 came Vic Steel, who would become director of research and development; Fred Adkins, who would rise to director of outsourced operations; and Greg Thompson, who would become vice president for sales. In 1994, we hired A.J. Nadler, who would become director of the silicon systems division, and Eric Creviston, Vic Steel's old college roommate, who would become corporate vice president for wireless products.

It is said that managers can be judged on how well they put in place people to succeed themselves who will be better than they, and we believed that. We closely watched the people we hired, talked at great length about who would be the better leaders among them, and singled out those for training and advancement.

As our hiring increased, it was these leaders we trusted to bring in people with the advanced skills, enthusiasm, and dedication that we demanded, and the congeniality that we desired.

If a company grows slowly, people are gradually absorbed into its culture. But a company adding people at breakneck speeds, as we were, faces dangers. People from work backgrounds with different ideas can establish beachheads that threaten the harmony of the whole. We knew that we needed to prevent that if we wanted to maintain the culture we had established.

Where once we had simply pointed people to their desks and put them to work, we now devised, under Liz Moore's direction, two days of orientation to explain our policies, our commitment to quality and ethics, and our sense of family within the company. We wanted every new employee to understand that hard work, imagination, friendliness, openness, and dedication would be rewarded, but that we had no tolerance for conniving, scheming, undercutting, and backstabbing. We wanted each to have a firm grasp of how things were to be done and what was expected personally and professionally. In the end, we wanted all of our employees to have the same excitement and pleasure in coming to work that we all still felt and treasured.

As a company, we prided ourselves on being able to react quickly to circumstances, particularly to customers' needs. Our management style always had been highly informal. In the early days, we often made decisions while passing in the hallways. Now many more decisions had to be made, many more people were involved, and everything was more complicated. Still we had to figure ways to make important decisions without losing our edge, our response time.

That couldn't be done through a big, top-down bureaucracy, and in the summer of 1999, we decided to break down the company according to product lines so we could put the decision making where the work was being done, and keep it fast and efficient.

We started with four divisions, one devoted to Nokia, another for power amplifiers, a third for digital cellular parts, and the fourth for silicon systems. At the end of the year we had added a fifth for broadband services, a division that would turn out to be short-lived.

Not only did this improve efficiency and allow us to keep our edge in decision making, it also helped to maintain the culture we had established. It was difficult to preserve cohesion in huge groups of employees, we believed, and we wanted to keep units below 200 people, so that everybody would know one another and feel a greater sense of belonging.

Results for the first quarter of fiscal 2001, which we announced on July 18, set new records, $98.1 million in revenue, with profit of $16.2 million, outperforming analysts' expectations. Dave predicted continued strong growth, and Jim Jubak, a market columnist for MSN, picked us as a "sure winner," declaring of our stock, "I really do think it will outperform Qualcomm over the next 12 months."

Despite the shakeout in the economy, our stock remained at $90, still considerably higher than it was at the beginning of the year. Back in the winter when it was climbing toward $200, reporters had begun speculating about whether we would declare another stock split, and we would have, if we could, but shareholders had not authorized enough new shares to allow that. We would have to wait until the shareholders meeting at the end of July, and late in June we had announced our intention to request another split.

At this meeting, before a packed ballroom at the airport Marriott, Dave reported that we'd had our best year ever, that profit was increasing faster than revenue, that we had grown to 1,000 employees and would be hiring many more. Our third fab was under construction, he said, and we were preparing to open our new testing lab on Gallimore Dairy Road, which was 10 times the size of our previous lab.

"We're preparing ourselves for what we believe will be a huge market in the future," he said.

We were focusing on new technologies, he pointed out, indicating that within five years we could reach $1 billion in annual sales. "But in this business," he added, "a five-year plan is good for about 20 minutes."

Our third two-for-one stock split, he announced to applause, would take effect on August 8.

After such a glowing report, wouldn't you know that the very next day, Nokia announced to everybody's surprise that it expected a slowdown in sales in the coming quarter, sending the stocks of chipmakers plummeting. Ours dipped to $64 before closing at $71.

Dave told reporters that we had seen no drop in orders from Nokia, and he thought the market reaction was overblown. "It's a short-term thing, if anything," he said, and we all actually believed that. "We'll just live through the temporary decline."

Our growth and development of new products required more and more money, and on August 15 we issued $300 million in convertible bonds to raise more cash. Our stock was trading then at $35.5, and these bonds could be converted to stock at $45.085 per share. They carried an interest rate of 3.75 percent. After discounts, we received $291.75 million, money that later would see us through some unexpected difficult months.

Later in the month, *Fortune* magazine released its list of the 100 fastest growing companies in the land, what it called "the speed-racers of the economy," and we were number two.

Number one was a New York company called 4Kids Entertainment, which held the rights to Pokemon.

We had continued expanding our reach during this year. In April we announced that we would be opening a sales and customer support center in Taipei, Taiwan, and in May another in Reading, England, our first in Europe. We were preparing to open a new design center in Pandrup, Denmark, in October. We also were planning to build our first plant outside the country, a packaging and testing facility outside Beijing, China, where several of our major customers already had operations. We would announce plans for the new plant early in October and expected to have it in operation within a year.

Our stock fared reasonably well after the split, closing at $35 on the day the price was halved. By early September, it had risen to $40, but it fell steadily over the next month, dropping to $23 on Tuesday, October 17, the day we issued our second-quarter results after the close of the market.

Revenue was terrific, another record, $102.2 million, with a profit of $17.7 million. But included in the report was a bit of information that would prove devastating. Our orders had been slowing over the past three months, and we expected a dip in revenue of perhaps 20 percent for the coming quarter.

Sales of cell phones had fallen unexpectedly, and inventory was backing up in manufacturers' warehouses. The long-anticipated new generation of phones also was being delayed, further reducing the need for parts. Still, our customers and market analysts were predicting continued growth in cell phones, and Dave told inquiring reporters that we expected this to be a short-term matter.

Only a few days earlier, *Forbes* magazine had named us number 41 on its list of the 200 best small companies in America, but that didn't stop investors from unloading our stock the day after our quarterly report. An astronomical 45 million shares changed hands, and the price at one point dropped by nearly 50 percent before finally closing at $14. Despite record sales and profit, and an unbelievable runup in the company's market value earlier in the year, the stock now had settled back to where it had been a year earlier.

Let me emphasize something here. The stock price obviously has been a major part of this chapter and the previous one. The primary reason for that is because it helps to illustrate what was going on in the economy at the time. But I don't want that to give a false impression. The stock price wasn't something that we obsessed about.

Oh, we exulted when the price soared, moaned and groaned when it fell. We didn't want our shareholders to be hurt by wild price swings, but there was little we could do to change that. It was apparent to all who cared to see that technology stocks were extremely volatile, and any investor had to be aware that it was as easy to lose money in such a market as it was to reap it.

Never did we sit around wringing our hands about the stock price or holding meetings about it. Our focus through all of this remained exactly what it had been from the beginning: to build the company in a conservative, honest, and responsible way. We knew that was how we best could build value for our shareholders, our customers, and ourselves.

Our big new fab, Fab Three, was completed in December, but we would not begin major operations there anytime soon. Sales and customer commitments to new products had continued to fall. We finally had brought Fab One up to full production of 60,000 wafers a year, and for now we didn't need the new fab.

Still we were moving ahead with transferring some of the technology, performing other functions, and hiring new people to work there. We remained hopeful of a turnaround and confident that we soon would be gearing up for production in this new facility, which we believed to be vital to our future.

Investors seemed to share our optimism about a turnaround in the economy. Our stock began coming back in November, and by early December, it was above the level it had been at the time of the split in

August. It dropped $10 by the end of the month, however, and closed the year at $27, down considerably from the same point the previous year.

In the nearly 10 months that had passed since the stock hit its high point, the market value of the company had fallen $12 billion. It was down $1.5 billion from the same date the previous year.

The new year didn't bring the upsurge for which we were hoping. Instead, it brought deepening recession, fewer sales, job losses, and company shutdowns not only across the country but throughout the world. One of our customers, Ericsson, which once had been the number-two cell phone maker and had 105,000 employees in 140 countries, had seen a drastic plunge in sales. At the end of January, it announced that it was leaving the handset business and selling its factories to a company in Singapore called Flextronics. Ericsson accounted for about 10 percent of our sales, but we already had a relationship with Flextronics, and we thought this move wouldn't hurt us much.

Of greater concern was Motorola, which once had been number one in cell phones. At our annual meeting the previous summer, Dave had told shareholders that it was going to be a close race in the coming year to determine whether Motorola or Qualcomm would become our second-biggest customer. Now it was clear that it wasn't going to be Motorola.

By the end of 2000, its share of the cell phone market had dropped to 13 percent, compared to 34 percent for Nokia, and in the first two months of 2001, it took drastic action, cutting nearly 10,000 jobs, closing or selling six plants, and preparing to eliminate four more. We definitely were hurting because of Motorola's problems.

And as events soon would prove, Qualcomm would end up being a disappointment to us as well. In February, Qualcomm had announced the new CDMA chipset of which our power amp was an essential part, and for which we were preparing to begin production. At the beginning of March, we expanded our agreement with Qualcomm to produce a power amp for a new module for cell phone base stations.

This announcement apparently alarmed some of our competitors, who feared us forming a major new alliance with Qualcomm. They launched a price war to thwart us, slashing prices on similar power amps to the point that we couldn't meet them without losing money, and we couldn't afford to do that. The matter would stretch out over the coming year until we and Qualcomm agreed to let our new alliance fade quietly away, a big disappointment to all of us, but especially to me.

At the time we were announcing the expansion with Qualcomm, things were not looking bright for us. A year earlier, our stock had hit its peak in the first week of March. But this year March began with the price dipping to $10, down nearly 90 percent in just 12 months. In only two months, our market value had dropped another $1.5 billion.

"RF Micro Suffers," read a headline in *The Business Journal* on March 9.

"Greensboro's 'Great New Economy Hope' is on the ropes..." the paper reported, more than a bit overdramatically, we thought.

The article went on to note that some of our competitors' stocks were faring better because they had wider customer and product bases.

One analyst, Jim Moeller of Dain Rauscher Wessels, pointed out that we were financially strong and well managed, but limited in scope for now, meaning that we still were too dependent on Nokia, something of which we were all too aware and on which we were working steadily and intently.

"We're in about as good a shape as anybody in the industry right now," Dave told Matt Evans, who wrote the article.

Of the stock price, he said: "We don't see that it's a fundamental issue....Most of us that have been around a while have seen this before. People around here just say, 'It'll go back up....'"

Only a week later Nokia, whose stock price had been cut in half by this time, announced that it expected its first-quarter profits to be in line with forecasts, despite slower sales. According to news reports, that cheered investors, but Nokia also made a prediction that didn't exactly cheer us. It said that global sales of cell phones likely would be 50 million fewer than expected this year. That sent our stock reeling again, dropping as low as $8.75 over the next couple of weeks.

April, however, brought not only showers and flowers but renewed optimism and, happily, stronger orders from Nokia.

Our stock began rising on April 5, and a week later it jumped more than $4 to $18, after we increased our revenue predictions for the first quarter of fiscal 2002. This came despite an announcement in March that revenue for our fourth quarter, which would end on March 31, would be considerably lower than forecasts. The new announcement prompted analyst Mark McKechnie of Bank of America to raise our rating to "strong buy."

After the market close on April 17, we released our quarterly report, which showed that revenue had dropped to $55 million, and we had suffered a loss of $6.8 million, our first in nearly five years. Moreover, while we predicted an increase in revenue in the coming quarter, we also expected another loss, though we anticipated returning to profitability late in the calendar year.

On a more positive note, we had closed fiscal year 2001 with revenue of $335.4 million, an increase of more than $66 million over the previous year, despite the wild swings and setbacks that we had endured. Our profit was about $35 million, down $15 million from the previous year.

Still we had managed to grow in sales as well as in facilities and people.

Investors responded positively to the report. Our stock jumped $6 the following day to close at $24.

At the beginning of 2000, after I had successfully negotiated the expansion of our licensing agreement with TRW, I was given a new title, executive vice president of marketing and strategic development. For years, I had been working to diversify our customer base and product line, but now, as these difficult times were proving, that necessity had become even more important. Finding ways to make that happen had become my primary focus.

We had to keep ahead of all new technologies, and to do that, we were getting ready to launch a new research and development department, to be directed by Vic Steel. We had assigned 20 engineers to it, and intended to add more.

We spent a lot of time talking about how to develop these technologies and harness them into practical products. When we saw products and technologies getting ready to present themselves that we wouldn't have time to develop ourselves, our alternative was to look at the possibility of buying the companies that had them, or creating partnerships to get them.

In the past few years, I had spent a lot of time looking at opportunities in other companies, but so far we had neither bought nor invested in any. We had, however, come close in 1999.

This was a profitable Florida company that had a plant in Costa Rica making filters to prevent interference in radio devices. I had gone to Costa Rica to inspect the plant and meet the employees—had some memorably great coffee while I was there.

We had hammered out an agreement to buy the company, had the lawyers involved and the approval of both boards, when I found out that a new technology was in the pipeline that would make this company's filters obsolete. We backed out at the last minute.

As it happened, this new technology didn't pan out. One of our competitors snatched up the company, and has since made a lot of money from it. If nobody ever has said that the technology business is unpredictable, risky, and sometimes downright confounding, I'll be happy to be the first to make that claim.

One thing had been certain to us for sometime, however, and that was this: if we were going to bring silicon up to half of our business we had to have a silicon facility of our own.

Our customers needed a stable, guaranteed source of supply and wanted us to be able to develop silicon modules so they didn't have to depend on standard parts. More importantly, they wanted them cheaper than we could get them made at IBM.

Early in 2000, I went to IBM to talk about the problems and possibilities. It would be to our advantage to work with IBM, we thought, since we had been using its processes for years. IBM was receptive, and we began a series of talks that over a period of months led nowhere, primarily due to resistance from top IBM executives.

Meanwhile, I had begun discussions with another company, Agere Systems in Allentown, Pennsylvania. Agere had been the micro-electronics division of Lucent Technologies, the former Bell Labs, but Lucent recently had spun it off as a separate entity. Agere was far bigger than we, with nearly $5 billion in revenue and more than 17,000 employees.

It also had all the silicon technologies and made every kind of chip except for radio frequency, and we had been providing some parts to Agere for a couple of years. If we could make an arrangement with Agere, we could design our parts to work with Agere's and provide a complete system for telephone manufacturers.

Like many technology companies after the bottom fell out of the market in 2000, Agere was in trouble, losing money. It had closed one foundry in Spain and eliminated 4,000 jobs throughout its operations. The company had a fab in Orlando, Florida, that wasn't running anywhere close to capacity. Nearly 400 employees had been laid off there.

This fab offered possibilities for us. Bill and I made several trips to Orlando to inspect it and meet with the remaining employees. With this operation, we could have a stable source of supply to satisfy our customers and reduce prices sufficiently to allow us to greatly expand our silicon business.

Over a period of weeks, we worked out an alliance that we announced in May. We would invest $58 million with Agere to upgrade the clean rooms and install new equipment in the Orlando fab over two years. Our products would receive first and unlimited access, and we would get them at a substantial discount.

We completed the deal early in July and made an initial investment of $16 million. We believed this to be a major step toward our goal of diversification, and we trumpeted it as loudly as we could. I told reporters that it was the biggest announcement we had made since our licensing agreement with TRW.

As promising as this arrangement seemed, we were by no means finished with expansion. We had several other deals in the works as well.

Our first-quarter results were charitably called "mixed." We had improved revenue from the last quarter, but fell considerably short from the same quarter for the previous year, $70 million as compared to more than $98 million in fiscal 2001. Even worse, we posted a loss of more than $33 million, $21 million of which was a special charge for inventory reserves and asset impairment.

Perhaps we could have avoided losses if we had taken the route of other companies in this tough period and laid off people. But that was something we had vowed never to do unless it became absolutely necessary to save the company. Some months earlier, I had made a presentation to the board about our diversification efforts in which I had stressed that our employees were our primary asset. We had worked hard to get really good people, and we were dependent on them and wanted to keep them. Laying people off is shortsighted and can damage a company far greater than it can help. Besides, our employees were loyal to us in good times, and we owed them the same loyalty in bad times.

We saw layoffs at other companies as opportunities for us. We had continued hiring and were getting good people from these companies. We now were bringing Fab Three online and were up to 1,300 employees. When better conditions came—and we could see that they were on the way—we would have people in place and be ready to leap ahead of less far-sighted competitors.

At our shareholders meeting on July 24, we kept an optimistic note. Dave reported that modules containing chips and other elements to provide certain functions were becoming our future. They had grown from 1 percent of sales at the beginning of fiscal 2001 to 32 percent at the end. And we expected them to account for 50 percent of our business in the coming quarter.

The recession seemed to be receding by early September, just as we expected. Orders had been increasing, and it looked as if that trend would continue. Production was up. In July, we had predicted sales of about $77 million and another loss for the second quarter. Now it was clear that we would do considerably better than that.

On the morning of September 10, we released revised projections for the quarter. We now expected about $91 million in revenue, and thought that we at least would break even. We also predicted revenue of more than $91 million for the third quarter. That morning the stock jumped more than $2, rising above $25, before closing near $23.

I was sitting at my desk the next morning when somebody—I can't remember who, maybe Kathy—poked a head into my office and said that a plane had hit one of the towers of the World Trade Center. Probably a small plane, I thought. Something must have happened to the pilot to cause the crash, a heart attack perhaps. Nothing more than a tragic curiosity. I continued with my work.

About 20 minutes later, Bill came into my office with a quick pace and a solemn face. Another plane had hit the second tower at the World Trade Center, he told me, a big jet. This could be no accident. It had to be a terrorist attack. We hurried to the board room, tuned the big teleconferencing screen to one of the cable news channels, and watched stupefied at the horrors unfolding in lower Manhattan.

The room filled quickly as the shocking news continued to flow in fast succession. Two hijacked Boeing 767s had been flown into the trade center towers. And at 9:37 an American Airlines Boeing 757 tore into the western side of the Pentagon. U.S. airspace was ordered cleared. The White House and the Capitol were evacuated, as was the United Nations complex in New York. We watched aghast as the southern tower of the trade center collapsed seconds before 10. There were gasps and tears in the room. Shortly afterward came reports that another 757 had crashed in Pennsylvania. Then just 29 minutes after the first tower collapsed, the second fell.

The people in these buildings had come to work this morning just as we had, expecting nothing more than dealing with the business of the day, and big jets had been deliberately flown right through their office windows, leaving untold thousands dead. The implications of these events were almost too profound to imagine. Clearly, we now were at war with a truly evil and evasive enemy, and life might never be the same again.

Our first reaction to all of this was purely personal. On any given day, we have a hundred or more people in the air. Could some of our employees have been on those planes, or on business in lower Manhattan that morning? As soon as the flight numbers became known, we got the travel office to check and were relieved to learn that all of our people were on other flights, in other locations.

Commercial flights would remain grounded for four days, and we had people stranded all around the country. We later got permission from the FAA to send a private jet on a one-day circuit to pick them up and bring them home.

Not for days did we begin to contemplate what all of this might mean for the company. Surely, it would have drastic effect on stock markets worldwide, and if they crashed, the global economy—70 percent of our sales were international—might fall into a tailspin and times could be even worse than those we had just gone through. There was nothing we could do about that, though, except go on with business as usual, and that was what we did.

U.S. stock markets remained closed until September 17, when the expected occurred. The Dow fell 716 points during the day, but closed down 679, the greatest one-day drop in history. Airlines and insurance stocks took near deadly hits. Defense and technology stocks fared better. Although the Nasdaq dropped 116 points, our stock, curiously, rose slightly.

The week continued its devastating course, ending with the Dow down more than 1,300 points, its biggest one-week drop. Our stock held its level through the week, but dropped to just under $20 on the final day.

Our second-quarter results proved far better than anticipated, more than $98 million, and we not only broke even, we returned to profitability a couple of months before we expected. It was a modest profit, just over $1.5 million, but still a profit.

And the next day our stock dropped $6 to $18, after some analysts lowered their ratings because of predicted slow sales of cell phones over the coming holidays.

We went right on with our expansion plans. In October, we acquired our first company outright.

RF Nitro Communications was a small start-up company in Charlotte that had been founded at the beginning of 2000 to attempt to commercialize a new semiconductor technology, gallium nitride, developed at Cornell University. The company had just a few people, and it was headed by Jeff Shealey, whose brother, John, had worked on the technology at Cornell and was serving as a technical advisor.

Dave and I drove to Charlotte, met Jeff, and looked over the operation soon after it got started. RF Nitro had gotten some seed money from VCs, had built a small clean room, and was working on potential products. Several companies had been trying to develop gallium nitride, but it had a lot of technical problems and none had made much progress.

The benefit of gallium nitride was that it could handle more power. Whereas silicon broke down at about five volts, and gallium arsenide at about 18, gallium nitride could take as much as 60 volts. It could produce more power in a smaller area more efficiently and operate at high frequencies, which made it a perfect material for cell phone base stations.

In the summer of 2001, RF Nitro produced the first transistors in gallium nitride and sent samples to us. The R&D department looked them over and passed them on to Bill, who thought they showed great promise. This was a technology we should have, Bill said, and late in the summer, we began negotiations with Jeff. By October, we had agreed that RF Nitro would become a division of RF Micro.

Dave explained the move to Matt Evans of The Business Journal, "It's a diversification from the handsets. When you live and die based on the ups and downs of one particular market, the volatility can just go crazy. Having a couple of different markets kind of levels that out."

This wasn't the only branching out that we were doing. We also had formed a partnership with a company in San Jose, Atmel, to begin jointly making parts for wireless local area networks (WLAN), which connect computers and other equipment by radio signals. These networks were becoming more and more popular, and lots of laptop computers were equipped with WLAN chips, making access to the Internet possible without a telephone or cable connection.

We had a high-flying prospect in the works, as well. Satellites circled the planet making it possible to precisely locate and track any person or object equipped with a proper transmitter and receiver. This was called the Global Positioning System (GPS). It was developed in the 1970s for military and intelligence agencies, but it since had been commercialized and promised to become a burgeoning business.

Shipping companies with huge fleets already were using it to track their vehicles. Some car manufacturers were installing GPS boxes in luxury models that would make it possible to find them quickly if they were stolen or endangered. GPS offered many other potential uses, but of particular interest to us was that the federal government had mandated that starting in 2005 all cell phones had to come equipped with GPS devices that would be activated whenever 911 was dialed, making it possible for authorities to locate any person reporting an emergency with a cell phone.

We had known for some time that this was a business we needed to be in and had been looking for access. We found it at IBM, which had been the first company to make GPS parts in silicon germanium, making them smaller, quieter, and requiring less power. IBM's GPS operation was in Irvine, California, and over several months I made many trips there and to IBM headquarters in Fishkill, New York, negotiating the acquisition of this unit. We reached agreement at the end of November and announced it early in December, once again to favorable reaction from analysts.

After the third stock split in August of 2000, TRW still owned nearly 24 million shares of our stock. It had agreed to hold at least a 10 percent interest in the company until May 2001, and it kept its word. But early in July, TRW started selling big-time, and it continued until the end of the year, shedding nearly 16 million shares. We didn't throw any parties when TRW's percentage in the company got low enough—less than 5 percent—that the sales couldn't drastically affect us anymore. But it was a relief, another uncertainty that now was eliminated.

We calculated that TRW had taken in at least $1 billion from our stock, not a bad return on the $25 million it invested. TRW had developed gallium arsenide HBT at a cost of $700 million, $400 million of which had been put up by taxpayers. Our relationship with TRW had repaid all those costs and still provided a handsome profit.

A couple of years later, a retiring TRW executive would tell me that TRW's investment in RF Micro had saved the company. The sale of our stock, he said, had allowed TRW to service its debt on bonds that might otherwise have been downgraded to junk status, destroying the company's credit rating. If that had happened, it would have been highly unlikely that TRW would have merged with Northrup Grumman, as it did in 2002.

True or not, this still goes to show how important it was that a handful of people at TRW were willing to listen when they got a call from a representative of a small company with no money and only two customers, asking for their help in creating a chip that nobody had attempted, and brashly telling them that we could take their technology and make them a lot of money.

RF Micro would close out 2001 with a market value of $3.3 billion, $600 million less than at the beginning of the year, but we felt certain that our strategic moves during the year had made us not only bigger, but stronger than ever.

Nonetheless, unpredictable outside forces—the continuing economic effects of 9/11, as that day of infamy now was known; threats of new terrorism; and the invasion of Afghanistan—would continue to plague us. And the aftershocks of the dot.com bubble weren't finished with us either.

12
Welcome to Moscow

The year of corporate scandal. That was what 2002 would turn out to be, a dismal time for business, integrity, and trust, a year that would linger in the public conscience and haunt Wall Street for years to come.

It had begun late in the fall of 2001 with the unraveling of the now infamous Enron, the Texas energy trading company where executives used complex machinations to disguise debt, create false profits, boost the stock price, and enrich themselves.

Enron's collapse cost employees their jobs and retirement savings, shareholders their investments, and brought down a major accounting firm, Arthur Andersen.

I was surprised that a huge company such as Enron could be involved in the fraud and deceit that was required to bring this about, because so many people would have had to know about it. Didn't anybody question these actions? Did nobody say, "Wait a minute, this is wrong and I won't be part of it.?"

I asked everybody on our management team if in previous jobs they ever had been in a meeting where somebody said, "Let's see if we can alter this report or shade this data to make us look better." None had, and none knew anybody who had been in such a situation.

If any such suggestion ever had been made at our company, that person would have been on the way out the door before he or she had finished speaking, but we knew that it simply wouldn't happen.

That's why in the beginning I thought that Enron had to be an isolated incident, and that the news media no doubt were blowing it out of proportion. After all, you'll be right most of the time if you suspect the press is blowing something out of proportion.

But 2002 brought corporate scandal after corporate scandal—Global Crossing, Qwest Communications, Worldcom, Adelphia, Tyco, Bristol-Myers Squibb, Halliburton, AOL Time-Warner, and on and on. Star analysts were revealed as touting stocks that they privately derided to increase profits for the banking and securities companies that employed them. And as if nothing were sacred anymore, Martha Stewart would be brought up on charges of obstruction of justice for lying about possible insider trading and eventually sentenced to prison.

All of these revelations left no doubt that there was rot at the heart of our system, and it was a sickening, even frightening prospect to digest. The cause was simple enough: a basic failure of character and integrity in the people who were allowing these things to happen. It could be seen as an extension of the greed and dishonesty that had caused the dot.com bubble to develop in the first place.

But there could be no doubt about the seriousness of the situation. Harvey Pitt, chairman of the Securities and Exchange Commission, summed it up later in the year: "We're dealing with issues and problems that will determine whether our capitalist system will survive."

From the time we began to realize the depth of the corruption, we knew that the consequences would be severe and that the innocent would pay along with the guilty. Clearly, the trust of the public—and of investors—would be lost, and like all forfeited trust, it would not easily be regained. The already repressed economy and stock markets were certain to become even more depressed, and everybody would suffer as a result. But there would be other consequences as well.

Cries for reform rose from every quarter, particularly from politicians and the news media (which had played a major role in creating the dot.com bubble), and as necessary as reforms surely were, we feared that in the white-hot heat of the moment, the cure might well be as drastic as the disease, especially for the great bulk of honest companies that were defrauding nobody.

The result was a rushed effort by Congress to legislate corporate ethics, the Sarbanes-Oxley Act, a lengthy and complex bill that was signed into law on July 30, 2002, by President George W. Bush. The president described its intention to "deter and punish corporate and accounting fraud and corruption, ensure justice for wrongdoers, and protect the interests of workers and shareholders."

All of those intentions are honorable and good, and maybe the bill will accomplish them. But it also created a public-company accounting board and vast new layers of overlapping bureaucracy to deal with untold numbers of lengthy reports that are being required from all companies, the honest as well as the dishonest, at great costs in time and money, to prove that they are innocent of the crimes of the few.

Companies of our size had to be in compliance with many of these provisions by June 2004, and it cost us several hundred thousand dollars to implement the process, a loss that our shareholders have to bear, although neither they, nor we, ever did anything wrong. The increased costs of auditing and report filing are huge and will continue into perpetuity. The burden is bad enough on trustworthy companies of whatever size, but it will be especially great on new and small companies.

If we had waited until this period to start our business, for this and other reasons (VCs, for example, were at this time fleeing from tech start-ups), we might not have been able to do it.

Amid the clamor for economic reform that arose early in 2002, politicians and the news media were quick to latch onto a handy cause to blame for this moral crisis: stock options.

Corporate executives were heaping stock options upon themselves, the argument went, then fraudulently manipulating revenues and profits to keep the stock price high for their own gains. Critics pointed out that top executives at Enron had cashed in more than $200 million in options before the company's collapse.

"The stock option culture," the *New York Times* said in an editorial, "is at the root of the current scandals."

No doubt, some executives did exactly as these critics claimed and should be punished for it, but that is no reason for demonizing stock options, and we rightly feared that was what was about to happen.

An argument can be made—and some have made it quite forcefully, most notably T.J. Rodgers, the outspoken CEO of Cypress Semiconductor in San Jose—that stock options built the high-tech industry and Silicon Valley. This is an industry that has created millions of jobs while other industries have been falling by the wayside, an industry that pumps more than a trillion dollars into the U.S. economy every year.

Small, innovative technology companies are expensive to start and always short on cash. The only way they can attract the talented people they need to allow them to thrive is by providing incentives through stock options. Why should an engineer, for example, leave a well-paying job at a large company to join a start-up operation that may not exist in a couple of years, if not for hope of ample reward? If the company does well, goes public, and the stock rises, everybody's happy.

This is true not only for new companies but also for established companies our size and larger.

High-tech companies, particularly semiconductor companies, are different from most companies. In our business, prices don't go up; they go down, often by 15–20 percent a year, as customers demand better parts at lower costs. The only way to make money is through innovation, ever-increasing productivity, and rapid expansion.

We have to keep attracting extremely smart, immensely talented, and highly motivated employees; people who are excited about their jobs, willing to put in whatever hours are required, and do it with enthusiasm. The way we do that is with stock options, which give them incentive to work hard and to stay with us. If we fare well, they will, too.

Stock options are used by most corporations, but most use them only to reward upper management, and therein lies the potential for abuse by people who are, to use a politically correct phrase, integrity challenged. Most high-tech companies, on the other hand, grant stock options to all employees, as we have from the beginning, because they require so many people with such rare knowledge, talents, and skills.

Stock options have worked particularly well for us, not only by offering employees a stake in the company's success, but by encouraging them to stay. We have one of the lowest employee turnover rates in the industry, nearly four times lower than the average for tech companies.

In most cases, the options we grant have a 10-year period during which they can the exercised. The option price usually is that of the stock price on the day of issue. These options don't become fully vested until after four years. At the end of one year, an employee can exercise 25 percent of the option, another 25 percent the following year, and so on. Each year, employees are reviewed and new options are granted on merit.

I should point out that there is nothing surreptitious about these options. The bulk of them go to rank-and-file employees, not to top managers, although they, too, are well rewarded. They are distributed according to the employee's contributions to the company's success. These options are approved by vote of the shareholders, who are fully informed that they will dilute the value of their own shares. The shareholders are willing to grant them because they believe, or hope, that by rewarding workers, production will benefit, and the stock will rise and reward everybody.

In the early '90s, a movement began to require stock options to be expensed at the time of issue against a company's profit or loss.

This made little sense in the case of high-tech companies for several reasons. For one, there simply is no way to determine the value of an option, particularly for the long-term options of highly volatile tech companies. It requires predicting the future, and nobody yet has made a sound science of that. For another, it would eliminate profits for many tech companies, and greatly reduce them for all, with the inevitable result of lowering stock prices and bringing about all the troubles that entails.

In many cases, these options could be worth absolutely nothing. If a company's stock price fell—as all tech-company stocks did after the dot.com bubble burst—and never climbed back to the issuing price, or strike price, as it is known, the option would have no value. Yet, a false value would have been determined for it and deducted from the company's profits, thus giving an utterly inaccurate picture of the company's actual results.

Reason prevailed when this movement began in 1993, and it was beaten back. But like the grinning gophers in that boardwalk mallet game, it kept popping up to be batted down again. The corporate scandals early in 2002 brought it roaring back with a ferocity never seen before, all the gophers rising at once, grinning defiantly and gloatingly.

In March 2002, U.S. Senators Carl Levin of Michigan and John McCain of Arizona introduced a bill to mandate the expensing of all stock options. The tech industry objected strenuously, and the bill got nowhere.

Levin tried to attach the essence of the bill as an amendment to the fast-moving Sarbanes-Oxley Act, but it got cut from that, too. The reformers then turned to the Financial Accounting Standards Board (FASB), an independent, industry-funded group of seven people who set the accounting rules for public U.S. companies.

In the spring of 2003, after the FASB made clear its intention to require expensing of stock options, two members of the U.S. House of Representatives from California, Anna Eshoo and David Dreier, introduced a bill to delay the requirement for three years. Barbara Boxer of California and Michael Enzi of Wyoming introduced a similar bill in the Senate, calling for the Securities and Exchange Commission to undertake a study of the issue. The FASB accused Congress of improperly intervening in its work, an indication of the arrogance that grows when a few people are given drastic power without oversight.

In September 2003, in response to the uproar in Congress, the board postponed the matter, saying it would issue a proposal early in 2004. But in November, it reversed the postponement and voted to require that options be expensed beginning in 2005. The move prompted Senators Boxer and Enzi, along with Senators George Allen of Virginia and John Ensign and Harry Reid of Nevada to introduce a new, compromise bill. This one would require options of the top five executives of a company to be expensed, but not those granted to other employees. New and small companies would be exempted from expensing any options. A similar bill was introduced in the House, and on July 20, 2004, it was overwhelmingly approved. But the measure remained tied up by strong opposition in the Senate.

If the expensing of stock options becomes reality, the method for determining the value of those options is almost certain to be the Black-Scholes formula. This formula was created by economists Robert Merton and Myron Scholes based on earlier work by a deceased colleague, Fischer Black. Merton and Scholes received a Nobel Prize for it in 1997.

The formula is complex enough to confound Albert Einstein, if he still were with us, and all but impossible for the vast majority of mere mortals to understand.

It was designed to predict the value of short-term options traded on stock and commodities markets, not for the long-term options that tech companies grant to employees. It probably works well enough for predicting the value of options for companies with fairly stable stock prices, but it has proved to be utterly unreliable in the case of high-tech options.

High-tech companies are high-risk companies, living constantly on the edge. A new product, or a scientific breakthrough, can spur rapid and tremendous growth in a particular company, while bringing down others whose products are suddenly obsolete. That, among other things, makes the stocks of high-tech companies highly volatile and brings into question just how valuable their stock options may be. They could be worth a huge amount, a small amount, or nothing, and since the dot.com bubble burst many have indeed been worth nothing. The Black-Scholes model, however, invariably grants high values to these stocks. This means that huge amounts will have to be deducted from the companies' profits, whether they are realistic or not.

Not many tech companies can take such hits. Most either will have to drop stock options, depriving millions of American workers of bigger stakes in the companies for which they work, seriously damaging initiative, and making it far more difficult for the companies to attract good people. Or they will have to relinquish profits, holding back or repressing stock prices, making expansion difficult, and putting them at a serious disadvantage against foreign competitors.

Entrepreneurial, start-up tech companies have been the driving engine of our economy in the past quarter century, and they will be seriously hurt if these rules are imposed.

A number of companies, Coca-Cola and General Electric prominent among them, have willingly begun expensing stock options. But these are companies that don't issue options to as many employees, percentage-wise, as high-tech companies. Their stock prices are fairly stable and not as overvalued by Black-Scholes. Thus they have far less to lose.

Nothing about the stock options that tech companies grant to their employees is dishonest or misleading. Nobody is being cheated by them, not the shareholders who approve them, not the IRS which still collects on them when they are exercised, or anybody else. Indeed, they have been the creative force behind our economy, and the FASB seems eager to destroy that for no better reason than to give the illusion of doing something to stop corporate corruption

"FASB has the green eye shades turned into blinders," Senator Boxer has said. "They are acting in a vacuum, as if what they are doing has no impact."

That impact eventually may be monumental to the economy.

As all the turmoil about corporate corruption swirled around us, we forged ahead with business as usual. Our stock had begun the year 2002 at about $20, and over the coming five months, as the scandals grew deeper, the index fell and our stock drifted downward with it to about $16. We, however, had started this year with big hopes, finally beginning production in Fab Three, which had stood largely idle in the year since it had been completed, due to the drastic downturn in orders.

Before January had passed, though, we had suffered yet another setback. Without advance notice to us, Agere put up for sale its Orlando fab, in which we had agreed to invest $58 million to gain a firm and affordable silicon source. We were stunned. This move defeated our purpose in investing in the fab. We already had put $16 million in new equipment into it, but suddenly our customers no longer could depend on us having a stable source of supply for silicon parts. Who knew who might buy the fab, or how long it might take, or what might become of our agreement?

Agere was unapologetic. It was a move that the recession had forced it to make, its executives maintained. They made no offer to return our $16 million, and we immediately put a hold on any additional money.

We were stuck with trying to protect our investment while renewing efforts to attain our original goal.

My first idea was to try to get IBM, which had provided our silicon parts for years, to buy this fab and dedicate it to our products. I arranged a meeting between representatives of IBM and Agere to talk about it, but weeks passed and nothing came from it.

My second idea was to explore the possibility of buying the fab ourselves. Agere was all too eager to accommodate us. I made numerous trips to Orlando to talk with fab employees and to Pennsylvania to meet with Agere executives. Bill went with me on several of these trips, and we both became convinced that buying the fab was what we should do.

In this depressed market, the fab, of course, was losing money, which was why Agere wanted to sell it. In fact, Agere practically was willing to give it away. I figured we could get it for a relative pittance, maybe $50 million, and it soon became clear that likely would be possible. Even with the additional investment for new equipment, we still would end up getting a fab worth hundreds of millions with a trained and dedicated staff for a fraction of its worth. A once-in-a-lifetime opportunity, it seemed to me.

Yet there was that distracting complication—the fab's outgo was so much greater than its income—and keeping it in operation until the markets changed and we could make it pay its way would be expensive. Bill and I thought the risk was worth it, because the return would be tremendous if it worked. And if it didn't work and produced the worst possible scenario—losing everything we put into it—it would be a big setback, but it wouldn't endanger the company. We never would recommend anything that could destroy the company.

I prepared a proposal for the board saying that the fab was a bargain that would solve our major diversification problem. I wouldn't say that this presentation got a cool reception. Icy might not even come close to describing it.

The board reacted as if I'd proposed hanging millstones around their necks and taking them for a swim. And it wasn't just the board. Others in our management team weren't exactly enthusiastic about it either. Only Bill and I were strong proponents, and Powell was willing to go along with us.

As I kept researching, refining, and resubmitting the proposal, opposition grew stronger. The board saw this as a far greater gamble than our first fab. Their fear was that it was a money pit with no prospects in sight to turn it around. And Bill and I realized that it was pointless to push it any further.

Fortunately, we had an alternative in the wings.

Agere didn't provide our only surprise early in 2002. In February, we got another one. Dave Norbury told us that he wanted to retire in the summer. He'd already informed the board. Dave was just 51, but he had been working long and hard for 10 years, and before old age started settling in, he wanted to take life a little easier, spend more time with his family, and follow other passions that he'd been neglecting—woodworking, primary among them.

We understood, but we didn't want him to go. Dave's leadership was all that we could have asked. None of our fears about an outside CEO ever proved true with Dave. He effectively presented management's views on to the board, and always did what he thought to be in the company's best interests. He had contributed so much to building the company, had represented it so well to the financial community and the public, and we had such a harmonious and smoothly functioning team that we hated to see it changing. The board asked Dave to stay at least until the end of the year to give us more time to prepare for the transition, and he agreed, to our gratitude and relief.

At the same time that we were getting word of Dave's intended retirement, we were opening a new sales and customer support center in Seoul, South Korea. Our Korean business had grown substantially. We had several major customers there, but the biggest was Samsung, a cell-phone maker that was growing quickly and challenging other major manufacturers, and we needed to have a close focus on it. Our sales in Asia now made up about half of our revenues, and later in the year we would open a similar office in Tokyo.

The recession still had our region in an unremitting grip that spring, but we were beginning to do better. Our fourth-quarter revenue was nearly double that of the same quarter the previous year at $100.43 million. We had only a modest profit of $2.77 million, compared to a much bigger loss for the quarter a year earlier.

Despite the hard times, we had increased sales during fiscal 2002 with revenue at $369.3 million, up nearly $35 million. We would have made a profit of about $2 million for the year if we hadn't taken write-offs for outdated inventory, for costs related to shifting to module production, for closing our production packaging line in Greensboro, and for opening the new research and development facility. Instead, we posted a loss of $20.58 million, our first annual loss since 1995.

We made news within the industry in May, when Dave chose a technology conference in San Francisco as the platform to announce that we were about to begin making our first prototype chips in gallium nitride. He called it a "disruptive technology" that would change the way things were done, and do for cell phone base stations what gallium arsenide HBT had done for handsets, greatly increasing output and efficiency. He predicted that we would begin production runs by 2004.

June brought another unsettling development. Some customers were delaying deliveries, and we had to cut revenue expectations for the first quarter from $107–110 million to $98–101 million (actually it would be almost $104 million). We expected a small profit (it would be $2.35 million). That still was a big increase from the same quarter for 2001, when we had $70 million in sales and a loss, but the stock fell $5.12 on the news, a 34 percent drop, closing at $10.12, a new 52-week low. It would drop much lower.

In July, Bill decided that it was time for him to step down as chairman of the board, although he would remain a member. The corporate scandals had drawn attention to board oversight, and Bill thought the board would be more independent without a founder and top executive as chairman. Al Paladino was elected as the new chairman.

The change was hailed in an editorial in *The Business Journal* as "a savvy, and investor friendly, move" that should be emulated by other area companies.

• • •

The answer to the problem created for us by Agere had come in a call in late spring from Theodore Zhu, vice president of sales, strategy, and business development at Jazz Semiconductor in Newport Beach, California. The management at Jazz had been aware of our situation with Agere while Bill and I were attempting to get the board to buy the Orlando foundry, which, incidentally, was never sold.

Jazz—yes, it's named for the music, suggested by an employee who was a big fan—was a new company. It had been the foundry for Conexant, a company with about 1,500 employees that created silicon systems for cable, satellite, digital video, and Internet connections. But Conexant had decided to become fabless and had spun out its foundry as a separate entity in March, although the deal hadn't been completed until May. Conexant itself had been spun out from Rockwell International in 1999.

Theodore assured us that Jazz, which had about 400 employees, was willing to do whatever was necessary to fill our silicon needs, even if it meant buying the Agere foundry. Jazz worked in advanced silicon processes, including silicon germanium, bipolar complementary metal oxide semiconductor (BiCMOS, pronounced buy—sea—moss), and radio frequency CMOS. BiCMOS has many applications with subtle technical advantages, is relatively inexpensive to produce, and is good for volume production. RF CMOS is fairly new and more complex than BiCMOS.

Theodore wanted Bill and me to come to Newport Beach for a presentation, and we agreed. We also invited Dean to go along.

The management team at Jazz put on an impressive presentation, laying out the framework of the deal we later would enter with them. We left confident that Jazz could fulfill our needs and that we could work well and amiably with the people there.

The Jazz arrangement turned out to be the easiest deal I ever handled. The board was so relieved that we were dropping the idea of buying the Agere fab that they practically welcomed the Jazz proposal with open arms. The deal essentially was this: We would invest $60 million in Jazz; Jazz would guarantee us low-cost production and work with us on developing new advanced silicon systems. We would acquire a minority ownership in Jazz (less than 20 percent) and get $60 million in discounts from the market price of wafers that Jazz produced for us. Part of the appeal to the board was that Jazz had great prospects for growth, which could make the investment quite lucrative.

We announced the deal on September 30, and I was appointed to the Jazz board of directors. Later, I was able to convince my old friend Don Schrock, who had retired from Qualcomm, to join the board as well. Few people know the semiconductor business as well as Don, and everybody at Jazz was excited about having his oversight and counsel.

• • •

We had held many long strategy sessions to put together a sound plan for diversification, and all of the executive staff agreed that we had to play a major role in every aspect of wireless communications. Jazz was the solution to just one part of the plan.

Our expansion into components for wireless local area network systems in 2001 had brought astounding success. We had shipped nearly $4 million worth of WLAN transceivers in fiscal 2002 and would increase that by more than 700 percent in fiscal 2003. We already had become the number two supplier of WLAN transceivers, but we wanted a much bigger profile in this rapidly growing field. The WLAN parts that we were making were used primarily for computer systems. They were for a set of specifications, or standard, called 802.11b. It allowed large amounts of data to be transmitted and received wirelessly, and it was the accepted standard at the time.

But other standards were being developed, and 802.11b eventually would be outdated. One of these was 802.11a, which would operate at higher frequencies and transmit data five times faster. It also would allow video transmissions, which we thought would give it great demand. Specifications for a third standard, 802.11g, had not yet been defined or released. It would operate at the same speed as 802.11a, and would allow video, but used the same frequencies as 802.11b. That band already was crowded, which could create troublesome interference for 802.11g.

We had to have 802.11a to remain competitive, and our engineers told us that we didn't have time to develop it in-house, as we had for 802.11b. We would have to form a partnership or buy a company to acquire it.

Our silicon systems division, headed by A.J. Nadler, was assigned to find out which companies we should be considering.

WLAN isn't the only technology for connecting devices wirelessly. Another is called Bluetooth. We wanted both.

While A.J. and his senior staff were visiting WLAN companies searching for the best prospect, I was looking for a Bluetooth partner. Bluetooth, like WLAN, is a short-range technology. It allows devices or equipment to be connected and controlled wirelessly. It could be used to connect a headset to a wireless phone, to connect a mouse or keyboard to a computer, or to send photos from a digital camera to a printer.

All of this was going on at the same time as I was working on the Agere problem and the Jazz deal. I discovered that a company in England was making Bluetooth components for Nokia. If it was good enough for Nokia, it probably was what we were looking for: a perfect match.

This company was CSR, Cambridge Silicon Radio. I made numerous trips to London, formed a relationship with the company's president, began talks about acquiring the company, brought the top managers to Greensboro to meet with our folks. Everything seemed right. I made a proposal to the board to buy the company, and they approved. I put together an offer, and CSR quickly turned it down. We never were able to come anywhere close on price, and I began looking for another Bluetooth company.

Meanwhile, our silicon systems division had settled on the best prospect for the WLAN technology we needed. That was Resonext, a company in San Jose with 90 employees, 70 of them engineers. Bill had gone out to see the company. So had Dean and Bob Bruggeworth, our new company president who formerly had been in charge of wireless products.

Everybody agreed that this was the company we needed, and I came in to negotiate the deal. Resonext, like us, had been funded by venture capital groups but at a far higher amount than we had received, about $70 million. Because of this, Al Paladino, our new board chairman, thought we'd have to pay a heavy price for it, especially considering that it had $30 million in the bank.

"If you can get it for less than $150 million, I'll give you a medal," he told me.

I ended up getting it for $133 million in stock, and we announced it on October 15, just two weeks after making known our investment in Jazz.

Dave heralded the acquisition as "huge" in our effort to broaden our customer and product bases. The WLAN market was expected to reach $1.5 billion annually by 2006, he said. Only a few big companies likely would be players in it, and we would be one of those.

At the same time, we released results for the second quarter. We had revenue of $119.7 million, profit of $6.9 million, a huge increase over the same quarter the previous year. But one analyst downgraded us because of the share dilution that the Resonext purchase would bring, and our stock dropped the following day from $8.66 to $6.91, a harbinger, alas, of things to come.

On November 14, the company sent out a news release that Dave would be retiring as CEO on January 10, although he would remain on the board of directors. Replacing him would be Bob Bruggeworth, our current president, who was 41, the same age Dave had been when he joined us. Dave had picked Bob as his successor, and the board had given unanimous approval.

Bob had joined the company in September 1999, after spending 16 years with Amp Inc., a supplier of electronic connection equipment. He had served as vice president of Asia Pacific operations and vice president of global computer and consumer electronics, based in Hong Kong, before coming to work for us as vice president of wireless products.

Bob had incredible energy from the moment he joined us, and it never has flagged. He can analyze a manufacturing situation quicker and more accurately than anybody I've ever encountered, and if there's a better organization and operations person anywhere, I have yet to meet him or her.

Equally important, he had fit harmoniously into our culture and management team from the beginning, just as Dave had done. We had no doubt that the company would be in good hands with Bob.

Several reporters did interviews with Dave about his 10-year tenure with the company. He told them that he felt good about retiring and that the company was "in the best shape it's ever been in."

"We've been fortunate that our hard work and intelligent risk-taking produced such tremendous growth," he told Richard Craver of the *High Point Enterprise*. "We're certainly a long ways from when our future competitors laughed at us...."

Dave said that he thought the greatest accomplishment during his time with the company had been building our first fab "when we had to stretch our necks a mile to bring off the financial commitment it took...."

"That was a nerve-wracking time," he recalled to Justin Catanoso of *The Business Journal*. "You'd wake up in the middle of the night and say, 'Okay, only 500 things can go wrong today.' At the same time, this place was so filled with optimism that we were downright giddy."

The company's astounding growth and the increasing responsibilities of managing such a far-flung operation had set him to thinking about retirement, Dave told Catanoso.

"I've always felt more at home in small- to medium-sized companies, and we had grown beyond that. I had to be honest with myself. I knew we had to find someone who is more comfortable managing at that level."

One thing he had especially enjoyed about working at RFMD, Dave said, was that "the management style here is so refreshing. Everyone has his specialty, and everyone trusts the other guy to do his job."

Bill once remarked to a reporter that when Dave came to work with us, we were like three guys digging a hole, and another guy comes along, picks up a shovel and joins in to get the job done. That was just how our relationship with Dave had been, how smoothly and comfortably we had come together to do the work and build the company. We were going to miss him, but we knew that he would find a lot of satisfaction in retirement

Dave told Amy Joyner of the *News & Record* that one of his financial advisers had said to him, "There are two kinds of retired people—the guy who wakes up and says, 'Oh, my God. What am I going to do?'...and the guy who wakes up and says, 'Oh, my God, what am I going to do first?'"

"I'm that second guy," said Dave.

• • •

One of the assets we got when we bought Resonext was a software division in Moscow—and not the Moscow in Idaho. We now had a dozen employees in Russia. Early in December, after the Resonext board approved the acquisition and the deal was completed, I went to Moscow to see the operation, welcome our new employees, and let them know what the company was all about. We'd hired a lawyer experienced in Russian labor rules to draw up the documents we needed them to sign.

I'd never been to Russia and didn't know what to expect, although I knew from watching *Dr. Zhivago* and other movies that in December it was going to be considerably icier and a whole lot colder than Greensboro.

My flight didn't arrive until after dark so I couldn't see much from the plane. Igor Grechkin, the administrative and finance manager of the office, was waiting at the airport with another employee to take me to the Best Eastern President Hotel. We'd hardly left the airport before we came upon a tall office building with flames leaping 30 feet out its windows. People were standing around looking at the spectacle, but not a single fire truck was in sight, and nobody seemed to be doing anything to control the fire.

"Welcome to Moscow!" Igor said, with a hardy laugh.

The hotel was nice, the food surprisingly excellent, with choices beyond beets and rutabagas, which I guess I had been conditioned to expect, not that there's anything wrong with those fine root crops.

After we finished work the following day, I still had several hours before my flight back to London, and Igor asked if I'd like to see the Kremlin and Red Square. I eagerly accepted, and he drove us there. It was even more impressive than it appears in photos and films. Igor parked so we could get out and tour for a while.

Before I left home, my daughter, Annette, also having seen *Dr. Zhivago*, had bought me a thermal hat and some Arctic-strength earmuffs to take along. I was getting them out of my briefcase, when Igor laughed and said, "You won't need those. It's not that cold today."

I took his word for it.

We got out and started walking into a brisk north wind, and after three minutes, I no longer could feel my ears. After five, I was wondering if they still were attached. Before we finished our tour, I was just hoping to get back to Greensboro with a few remaining fingers and toes.

A word of advice. If a Russian, no matter how well intentioned, tells you that it's not that cold today, just remember that Russians have spent most of their lives living in a deep freezer and have no real concept of temperature.

• • •

The company put on a big dinner for Dave, his family, friends, co-workers, and board members at his house on the evening of his last day at work, Friday, January 9. I was emcee for the program and lots of people said lots of nice things about Dave. I had been entrusted with the comedy for the evening, and if I do say so myself, since nobody else seemed willing, it was one of my better pieces of work.

I had prepared a slide show of photographs touching on some of the highlights of Dave's career at the company. Some of the photos, I must acknowledge, had been slightly altered, thanks to the amazing abilities of computers and our graphics people. Well, maybe more than slightly altered.

"We were deeply touched when Dave asked for a photograph of the three founders...," I intoned early in my bit, as a formal portrait of Bill, Powell, and me flashed on the screen.

"...until we came in one day and saw what his plans were for the photo."

On the screen came Dave at his desk, laughing wildly, Magic Marker in hand, drawing dunce caps, beards, mustaches, and Marx Brothers hairdos on the three of us.

"We had a series of groundbreakings, and at the first one Dave Vandervoet gave a speech about implanting cell phone electronics in a tooth. Dave tells Vandervoet that he will be the first to have this implant...."

Photo of the two Daves sitting side-by-side at the ceremony, Vandervoet holding a chip up to Dave's open mouth.

Quick cut to photo of the two Daves at the following reception, our Dave grinning, a black chip replacing his middle upper tooth, giving an impression of a gap bigger than David Letterman's.

"Unfortunately, Vandervoet chooses the wrong tooth."

It went on like that.

Dave actually might have been able to put a Letterman gap to good use. He loved to whistle, and it drove some people crazy at work. What a lot of people didn't know, I told the crowd, was that Dave had forced the executive staff to form a whistling chorus to accompany him, and there on the screen were all of us, posed on a staircase, whistling away. To this day, I'm told, some employees live in horror that this may be true and that we may come around performing at any moment. Management consultants may want to consider this as material for keeping employees in line.

Not long after Dave joined the company, he and I went to Portland, Oregon, to visit Triquint Semiconductor in nearby Hillsboro. As we were crossing one of Portland's many bridges over the Willamette River, Dave said, "Let's see if we can find Bridge City Tool Works."

Bridge City makes the world's finest woodworking hand tools. They are handcrafted of brass and rosewood and other high-quality materials in limited editions. Pricewise, or otherwise, you won't find anything to match them at Home Depot. When Dave ordered tools from Bridge City, he always bought two, one to use in his home shop and the other to keep pristine in its box as a collectible.

We drove down beneath the bridge into a rundown warehouse-industrial area and soon came upon a little sign that said "Bridge City Tool Works." We parked and climbed a flight of steps to a tiny office, where a woman sat behind a desk.

"We're here to do a factory tour," Dave announced.

"Fantastic," said the woman, who hurried off to fetch the plant manager. He came out beaming, introducing himself, shaking our hands and telling us how wonderful it was that we had come. They apparently don't get many tourists in this area.

We got the full treatment, and met the immensely skilled craftsmen who make these beautiful tools, many of whom seemed to favor pigtails. Dave was so delighted that I began to wonder if he might show up at work with a pigtail himself one day.

We had ordered a commemorative hand plane from Bridge City as a retirement gift for Dave, two of them actually, but this tool wouldn't be issued until April, and we had only a certificate to present him.

I was not without reward myself on this occasion, I might add. At the end of the program, our board chairman, Al Paladino, rose to say that when we had begun negotiating to buy Resonext, he had promised me a medal if I got it for less than $150 million.

He produced it with a flourish. It was a big thing, appeared to be made of bronze, with a fancy ribbon attached. I was impressed until I saw that it was for perfect attendance at the Knights of Columbus. Only later did I learn that Al had bought it at a flea market, sparing no expense. I cherish it nonetheless.

13
Revolution, Phase II

By the beginning of 2003, convergence had become the big, new buzz word in our business. It was the concept of bringing a lot of functions together in a single device. In other words, a cell phone would essentially become a pocket computer with multiple uses. It would have a color screen; a digital camera; high-fidelity stereo player; a global positioning satellite unit for navigation and help in emergencies; Bluetooth and WLAN chips to allow connections to computer networks, the Internet, printers, headsets, and other devices; and eventually high-resolution video receivers and transmitters, even TV reception.

Convergence summed up the future for us, and we intended to have a leading role in every aspect of it. But like everything else in technology, convergence presented a host of problems that had to be solved. Many of the different components that were being joined in one small unit conflicted and interfered with one another. That was one reason why these much touted third generation cell phones had been so long delayed.

We'd heard about a small company in Boulder, Colorado, called Channel Technology, which was designing complex digital systems to deal with the problems of convergence, and Bill and I visited the company several times. It had only five engineers, but they were brilliant at what they were doing, and we wanted them working with us. I negotiated a deal, got the board's approval, and in February we bought Channel Technology and turned it into another of our design centers.

We still were lacking Bluetooth technology. Back when I was trying to acquire CSR (Cambridge Silicon Radio) to get it, I also had been considering another company, Silicon Wave in San Diego. It had been founded in 1997 and had become the first company to deliver a fully qualified single-chip Bluetooth radio modem.

I had spent a lot of time talking with Silicon Wave's president, Dave

Lyon, but CSR had seemed a more promising prospect because of its relationship with Nokia. After the CSR deal fell through, we put a Bluetooth acquisition on hold and continued trying to develop the technology ourselves. We since had spent about $11 million to that end without getting much from it.

Like many technology companies, Silicon Wave had faltered during the economic downturn and needed money. Early in 2003, Dave Lyon approached our board member Erik van der Kaay to ask if we still might be interested in acquiring the company, and we resumed talks.

A major investor in Silicon Wave was a Boston VC group headed by a friend of Al Paladino. Bill and I flew to Boston to meet with them and work out the framework of a deal. Instead of buying the company, we offered to invest in it in return for sole rights to manufacture and distribute its products. We announced the deal late in May.

Meanwhile, we continued our own climb out of the long decline that had followed the technology bust three years earlier. Revenue for fiscal 2003 hit a milestone, finally crossing the half-billion-dollar mark at $507.8 million, up from $369.3 million in 2002, substantial growth in anybody's book. Unfortunately, we posted our second annual loss in a row, $9.3 million, due to a write-off of nearly $19 million for expenses related to the purchase of Resonext and an interest rate swap. Otherwise, we would have had a profit of about $10 million, admittedly a small margin for revenue so great.

Our stock had been drifting downward with the rest of the market due in part to uncertainties about the war in Iraq, which had started on March 20, and a deadly new disease called severe acute respiratory syndrome (SARS), which had killed 500 people in Asia, 23 in Canada, and a handful in the U.S. We had been particularly affected by SARS because of our new plant in Beijing. We regularly had employees going back and forth to China and other Asian nations and had to stop all of that travel. We were overseeing the China plant by e-mail and telephone conferences.

News of our second annual loss caused the stock to drop below $5 before closing at $5.10. Eight days later, it dipped to a depressing new low, $4.55. As despondent as we were about that, we could take solace in this: that original $12 share of stock issued in 1997, adjusted for splits, still was worth $36.50 at that price, more than triple its value in less than five years, a solid return under normal circumstances, but little comfort to somebody who got caught up in the heady days of the dot.com bubble and bought at an inflated price.

We were keenly aware that if we wanted to improve the stock price we had to raise profit margins, and we had a plan for that. It included investing $40 million in new equipment to allow us to produce six-inch wafers, instead of the four-inch we had been using from the time we opened the first fab. Six-inch wafers were thinner and more fragile, thus susceptible to breakage, but they would give us twice as many chips and reduce manufacturing costs by 30 percent. We planned to begin production of these wafers in late summer.

Some of the money for this equipment would come from another bond issue. On July 1, we sold $230 million in new convertible bonds. Our stock was trading then at $5.78, and these bonds could be converted to stock at $7.63 per share. The interest rate was 1.5 percent. We planned to use $200 million of this money to redeem the bonds from the issue three yearts earlier at the much higher interest rate. (A year later, we would redeem the remainder of the bonds from that first issue.)

One thing that I've noticed about stock prices is that when they are down, people—and not necessarily all of them stockholders—seem to take pleasure in kicking a company around, whether it's deserved or not

We got kicked pretty hard in the *News & Record* two days before our annual shareholders meeting on Tuesday, July 22. The main headline across the top of the Sunday business section was this: "Dipping chips: RF Micro stock down." Actually, at this point the stock wasn't down as much as it had been a couple of months earlier, when it had received no such notice, but it was down far too far for sure.

"Investors have few kind words for RF Micro Devices these days," the article began.

"'This stock is dead.' 'Bail out.'..."

I had heard and read these comments, and worse, myself, and they stung. I had great empathy for people who had bought the stock high and seen it go low, but likewise there were people who bought low and had gotten out when the price was extremely high. Some had grown wealthy on modest investments in our stock. All a matter of timing— and maybe more than a little luck.

I also had heard people derisively dismiss the company as a failure because of the low stock price. That didn't just sting. It hurt to the core. As slow as I am to anger, that almost is enough to make my blood boil.

In just 12 years, we had created a company that changed an industry and provided jobs for more than 1,800 people—and that number was growing daily. We were bringing in more than half a billion dollars in revenue annually, contributing hugely to the area economy, and supporting a big chunk of government in Guilford County and North Carolina with our taxes. We had plants and offices in 22 cities in 13 countries. We had created products that no other company had conceived, products that made people's lives better, and we were working on many more. We now were the world's leading supplier of integrated circuit power amplifiers, and our market share was growing rapidly. We were constantly working on new products and building a solid company for the future.

That, it should go without saying, is not failure by any definition.

As it turned out, our annual meeting was quite amicable, without questions or objections by shareholders, who seemed to be impressed by Bob Bruggeworth's presentation about our long-term strategy that was leading us beyond cell phones into a great variety of new wireless products.

An hour after our shareholders meeting, Lehman Brothers in New York upgraded its rating on the company from neutral to positive, saying that the ebbing of the SARS outbreak and cell-phone manufacturers' inventories could mean better times for wireless chip makers.

We had been seeing signs for some time that business was about to improve dramatically. We were getting unprecedented orders, had begun hiring new people, and were gearing up for a tremendous upsurge in production. It appeared to us that the long hoped for turn-around in the economy might at last be upon us.

We were grateful that our board had the foresight to allow us to build Fab Three, because we wouldn't have been able to handle all of this business without it. Although Fab Three had sat largely idle for a year after it was completed, it was vital to our future. I had said all along that Fab Three was the best sales tool we had, because it gave customers confidence that we could meet whatever needs they might have, and that now was proving to be true. We were swamped with business.

Despite all of this good news, more than a few analysts remained down on the company. So widely divergent were analysts' opinions that *The Business Journal* devoted an article to the contrasts on July 28.

Analysts are sort of a touchy topic because nobody wants to risk offending them, so let me say that what follows is my opinion, not the company's.

I've heard that there's only one way to look at analysts. Those who say good things about your company are brilliant and to be heeded; those who don't are biased, or ignorant.

Let me say up-front that I have no doubt that the overwhelming majority of analysts are knowledgeable, conscientious, and make every effort to understand and do a good job. Most have been fair to us. But a handful seem to have...how should I put this? Closed minds seems a little harsh, but it may be apt.

Some appear to have favorites, and if you're not one of those, you're unlikely to get a fair assessment.

When our stock was booming, some of the same analysts who find regular fault with us now couldn't come up with enough good things to say. But once the stock collapsed, they couldn't find anything good to say no matter what we did or how positive the financial results. It was a reverse image of what happened when we were on the way up, a trap from which we couldn't escape and which we didn't realize was being set for us when things were looking so good.

Analysts criticized us for years because of our dependence on a single customer, Nokia, who provided as much as 85 percent of our revenue at times. But when we increased our product line and customer base to the point that Nokia was accounting for only 40 percent of revenue in 2003, some analysts claimed that our diversification efforts had been too scattergun and faced too much competition to matter.

If we posted good results and bright expectations, inevitably some analysts would claim that our customers were overloading on parts, giving us little chance for future sales. To some analysts the horizon was perpetually bleak.

Maybe this is a reaction to the boosterism of the dot.com bubble and the scandals involving some major analysts. If you see only gloom, you can't be accused of donning rose-colored glasses or touting stocks of companies that send banking fees your way. But constant negative views hurt companies as well as the economy.

To be fair, it seems that the most common thing that causes analysts to turn on a company is disappointment. If they predict something positive that doesn't come about, they get burned. They apparently think that makes them look bad, and they are far less likely to say good things again, even though their expectations oftentimes are simply not possible due to circumstances beyond their view, no matter how motivated or well-run a company may be. Perhaps the most troublesome trait about analysts in general is that they are more tactical than strategic. Their primary focus is the next 90 days, which is utterly meaningless unless a company is in a life-or-death situation, especially for tech companies, which have to spend huge amounts in research and development on time-consuming projects.

A company that's doing what's best for the long-term interests of shareholders isn't apt to be embraced by analysts. Lay off 30 percent of the work force, though, and that company will be hailed. That's expense cutting, and that's almost always good in the eyes of analysts. But whether such a short-term solution actually is helping or hurting a company is a far more complicated matter to discern.

One thing we've noticed since the market crash that began in 2000 is that even analysts who aren't overtly hostile to us usually underestimate our results whether good or bad, but especially when they are good.

That was true for the second quarter of fiscal 2004, which we reported in October 2003. We posted record sales of $163.5 million, with profit of $11.4 million, or six cents a share. Analysts had been predicting $145 million in revenue and no profit. Furthermore, we expected an even better third quarter with sales of $180 million and profit of eight cents a share. But some analysts just wouldn't accept that this was laudable.

By the time we reported these results, we were ready to make known a phenomenal new development. More than half of all the cell phones being made in the world now contained our power amplifiers. That was double the number of a year earlier, an incredible gain in market share in so short a period. We were by far the world's leading supplier. Our power amps were in nearly three fourths of the phones being made in Taiwan, a major manufacturing center, more than triple the number from just two years earlier.

This news, along with a rising market and our record second quarter, caused our stock to start inching upward after dwelling so long in single digits. By January 20, 2004, it had reached $12.26.

At the end of the market day, we announced our third quarter results, which were much better than we expected, another record, $193 million in revenue, with profit of $28.2 million, or 13 cents a share, outstripping not only analysts' expectations but our own.

Record sales and profits. Great news, right?

So why did the stock plummet in after-hours trading? More than 67 million shares exchanged hands the following day, and the price closed down $2.41 at $9.85. It would drift lower in following days.

This may have been because the results came with a word of caution. We were entering a transition period, a market shift. A whole new wave of cell phones was on the way, and sales of phones with the older technology would be dropping as customers became aware of the new ones. Also fourth quarter sales are normally lower than those in the third quarter. Still, we would be profitable and produce revenue much greater than for the same period the previous year, although we might not quite match analysts' expectations.

In this topsy-turvy market, though, it seemed that great success produced only fright and merited only punishment.

By the beginning of 2004, we had made two investments in Silicon

Wave for a total of $8 million. The Bluetooth products the company was developing were coming along well, but not yet providing revenue enough to meet their cash burn. Silicon Wave was running out of money again, and having trouble finding other investors, perhaps because we held exclusive rights to manufacturing and marketing.

We knew, too, that Bluetooth would be designed into future cell phones, and that the major manufacturers might not be willing to rely on a company in Silicon Wave's situation for the chips they needed. We certainly didn't want Silicon Wave turning to a competitor, so the reasonable solution was for us to acquire the company outright and provide money necessary to get these new and promising chips to market. I began negotiations to that end in January.

Because of our experience with Resonext, where products were developing more slowly than expected, our board was reluctant to make another huge outlay of stock and cash unless we could be certain of a timely return, so I chose a new course for Silicon Wave. The negotiations were predictably difficult, but by the end of February, we had essentially reached a deal, although it would take another couple of months to hash out the details.

We would put up $10.8 million in cash and tie additional payments to sales, an earnout, this is called. For 2005, we would pay .5 percent on sales above $6 million. For 2006, payments would match sales above $25 million, with an overall cap of $75 million. We hoped we'd have to pay the full $75 million, because if we did, this would be an extremely productive and valuable addition to our company.

Early in April, we and market analysts were surprised when Nokia announced that it expected a 2 percent decline in sales for the first quarter. Overall, cell phone sales were growing, and Nokia still was the top producer, but Motorola and Samsung were coming on strong, and customers were favoring designs of their new phones over Nokia's.

Nokia's stock dropped nearly 19 percent on the day after the announcement, and we expected a big blow as well. Nokia now accounted for only 35 percent of our revenue, down to the levels we had been striving to reach, but analysts still claimed that we were tied too closely. As it turned out, we didn't suffer quite the blow that we expected. The stock dropped from $8.90 to $8.27 on the day after the announcement, but climbed back to $8.71 the following day.

Our strategic plan to diversify and push us past the billion-dollar revenue mark presented the need to restructure the company so that we would be prepared to meet the demands that our coming sophisticated new chips, new customers, and increased production would bring.

On April 20, we announced that we would realign the company into three basic business units: cellular, wireless connectivity, and infrastructure. Cellular would handle the power amps and transceivers for phones, which made up the great bulk of our current business. Bluetooth, WLAN, and GPS parts, for which we expected incredible growth, would fall under the wireless connectivity unit. We just were beginning to design and make chips for cell phone base stations, employing a radical new technology, and that would be the focus of the infrastructure unit.

A week after announcing the restructuring, we released financial results for fiscal 2004, and once again they exceeded analysts' predictions. We had sales of nearly $651.4 million, up 28 percent from $507.8 million in 2003. Profit was $29.7 million, or $.15 per share, compared to a loss of $9.3 million, or $.05 per share for the previous year. Our stock dropped from $8.47 to $7.75 on the news, probably because we had posted a small loss of $859,000 for the last quarter. What can you say?

Spring had brought the beginning of what Federal Reserve Chairman Alan Greenspan later would call a "soft patch" in the economy, due, it was surmised, to rising fuel costs. By early summer, the soft patch had become a downhill slide for technology companies.

"The High-Tech Balloon is Going Sssssssssss," read a headline in The New York Times on July 11.

The following day, Merrill Lynch downgraded the entire semiconductor sector. Four days afterward, Nokia, which had been fighting to regain customers by cutting prices and rolling out a new line of phone designs, warned that it would show another drop in sales for the second quarter.

We had a strong balance sheet, with nearly $200 million in cash reserves, money we were intending to use in part for imminent expansion of production. We also were preparing to issue strong first-quarter results for fiscal 2005: sales of $165.77 million, up 21 percent from the same quarter a year earlier, with profit of $3.01 million. We released this news on July 20, and the next day three Wall Street firms downgraded our stock, citing the same old refrain—too close to Nokia. Our stock dropped 13 percent that day to $5.73.

A week later, with the stock again at a low ebb, our managers and directors presented our strong confidence in the company's future at the annual shareholders meeting. This time some shareholders did express their dismay over the stock price. Bob Bruggeworth told them that he understood their frustration but explained that part of the company's shortcomings in profit in the past couple of years had been due, in addition to the economic downturn, to the costs of diversifying and developing new products that soon would begin to pay off.

Even as this meeting was in progress, a major new development was taking place that confirmed what Bob had been telling the shareholders.

For three years, we had been working on a project called Polaris to create the most technically complex products we have ever produced. Polaris combines all of the transmission and receiver parts of a cell phone, with the exception of the power amp, into a single tiny unit. It will work with new broadcast standards such as EDGE, GPRS, and WCDMA that allow huge amounts of data to be moved and will make possible the new, multi-use cell phones created by convergence. Because it works better with our power amps, it will greatly increase the dollar amount of the parts that we get into cell phones.

Polaris I, which we had begun shipping in limited numbers in the spring, consisted of three chips. But most of our efforts had been going into the far more complicated Polaris II, which uses only two chips. Polaris III, on which we are working hard, will combine all of a cell phone's transmission and receiver functions in a single chip.

Motorola had designed Polaris II into a new series of phones which it planned to launch with great fanfare late in 2004, and we already had begun shipping chips for the prototypes. Polaris, incidentally, is based in silicon, and will be the long-awaited solution, I'm convinced, for all those schemes that I kept coming up with over the years to bring our silicon production up to the levels we were seeking. Because Motorola needed these chips in such large quantities, we were having them manufactured at TSMC in Taiwan, one of the world's largest semiconductor foundries.

Early in August, we heralded a banner event in typical fashion. We had just shipped our one billionth power amp. We pitched a big tent in the parking lot behind corporate headquarters and held another free lunch for the 1,500 employees in Greensboro who had worked so hard to bring this moment about. Lunch this time came from Prissy Polly's, a popular barbecue restaurant in nearby Kernersville. Most employees came wearing the bright green t-shirts we'd had made to commemorate the festive occasion.

I conducted a press conference to which we had invited many community leaders. It was held in an adjacent tent just before employees started lining up to eat. I told the story of our mysterious, disappearing first power amp, the 2103, that we agreed to make for Nippondenso in our first year in business. I've told it so many times in so many settings that I've worked it into a comedy routine that never fails to get laughs.

But the story has a serious point, of course.

Even after Nippondenso cancelled the contract and left us as deflated as one of those molten chips, we didn't give up. We saw failure as nothing more than opportunity, forged on, and fixed the problems. The 2103 had delivered our company to this proud point.

"The opportunity today," I told the group, "is even greater than it was in 1991 when we started."

When Bill and Powell incorporated RF Micro Devices, the revolution that cell phones would bring to electronic communications was just beginning. Three guys with little money in Greensboro, North Carolina—and I'm proud to be one of them—led that revolution by creating the first radio frequency integrated circuits and by commercializing an exotic technology that few people thought would work for that purpose. Not only is RF Micro Devices now the world's leading provider of power amplifiers, but the seventh largest manufacturer of semiconductors for wireless communications, a ranking that we intend to improve.

Although this book is coming to a close, this by no means is the end of the story, just the beginning of a bigger one. We are poised for the second phase of the cell phone revolution. This will be even greater and more significant than the first, opening amazing possibilities for us and for the users of the magical devices that are on the way.

Just as we were at the forefront at the beginning of the cell phone revolution, we now are in the same position for the second phase. Naive revolutionaries though the three of us were before, this time we are more than 2,200 strong around the world, and growing, with some of the best minds and most innovative ideas in the business. This time, too, we are far better organized and financed—and we know exactly what we're doing, because we've been planning and working toward it for years.

Polaris, our north star, will guide this new revolution. It will be powered by our power amps and strengthened by our Bluetooth, WLAN, and GPS chips, for which we anticipate an explosion of orders in the coming year. We now are refitting Fab One to start major production of the first high power amplifiers for cell phone base stations in gallium nitride, a new technology that will have the same effect on infrastructure as our gallium arsenide HBT power amps had on wireless phones, changing the way they operate, and opening a big new field of business to us.

February 2005, will mark the fourteenth anniversary of RF Micro Devices. I'm told that it's unusual for founders of a company of our size and age to remain so closely involved in its management, but it doesn't seem unusual to Bill, Powell, and me. Our dream has grown with the company, and it grows more exciting as we continue to build it. We still feel like kids who don't want to leave the ball field to come into the house at night. We're having too much fun.

Afterword

I love making talks about our company, and I do it regularly before all sorts of groups. I particularly enjoy speaking to students at university business schools and answering their questions.

Maybe it's just the hard economic times we've been through in recent years, and the realization that high-paying jobs won't automatically be waiting for new MBAs, but I find that ever-increasing numbers of these students are dreaming of starting their own businesses.

This is always encouraging to me, because I believe that new, entrepreneurial companies will continue to be the driving force of the economy.

These students are always brimming with ideas and enthusiasm, and I have to admit that I sometimes see my younger self in their faces and envy them the experiences that lie ahead.

They also are hungry for encouragement and advice and filled with questions—loads of questions. Sometimes their questions are as difficult and demanding as those Bill, Powell, and I faced from the VCs in our early attempts at money raising, and oftentimes I have no satisfactory answers.

Now and then, some of these budding entrepreneurs seem to want road maps that nobody can provide, guides they can follow from point to point and reach their eventual destinations without wrong turns, detours, or head-on collisions.

There are no magical routes, of course, no easy answers or guaranteed rules for success despite all of the how-to volumes that fill the shelves of book stores and libraries, no certainty even that good ideas, adequate capital, hard work, and determination will be enough.

I don't pretend to be an expert on starting and building companies, but from the experiences that Bill, Powell, Dave, and all of us survived, I have accumulated some general thoughts that may be helpful.

Optimism, as I've mentioned, is vital, but beyond that, I think a certain attitude is necessary.

I was on a panel for business students once at Wake Forest University in Winston-Salem, and after I had described all the long hours we put in, the frustrations and failures that we endured in the early days of the company, a young woman raised her hand and asked, "Is the success that you've obtained worth all that you had to give up to get it?"

I was taken aback. I never thought that I'd given up anything. If you loved golf, would it be a sacrifice to have to play 18, or even 36, holes every day?

If you think of work as sacrifice, then you probably shouldn't start a business.

I know that there are people who think that work is a necessary but mostly unpleasant experience that is to be minimized as much as possible on the way to early retirement. People who think that way should get 9-to-5 jobs, or less, if they can get away with it.

If you're going to start a business, work shouldn't even enter your mind. If you have a passion for what you want to achieve, the effort you put into it will be like golf to the person who loves golf.

I know that some will reject the idea of work ever being fun, but, believe me, a passion for what you're doing makes it that.

The very idea of sacrifice can be harmful to the entrepreneur. If you're constantly thinking of what you may be giving up instead of for what you are reaching, it can hurt you in many ways. If, for example, we had worried to the point of indecision about the percentage of ownership that we were handing over to the VCs to get the money we needed, we might never have gotten the company going, or kept it growing.

Another thing that's vital is this: If you're thinking about starting a company with the idea of getting wealthy, then you have the wrong motivation. Such misdirected focus is likely to greatly diminish your chances. Money should be a byproduct of what you're doing—a pleasing one, for sure—but never a goal.

Now to the practical. Once you have a firm grip on the products or services your company intends to offer, the first thing you need to do is research and write a plan for the business.

I'm not talking about a quickly assembled, pie-in-the-sky, run-of-the-mill business plan. I mean a really thoroughly researched plan, the kind that Al and his partners required of us after we thought we had been thorough; a complete, in-depth analysis about markets, costs, profits, competitors, capital, and all of the other details that will be essential to knowing and understanding what you are planning to do.

This not only will give you a chance to see the flaws in your ideas and plans, but it will give you a sample of the discipline that will be required to bring them about. That you're doing all of this difficult and time-consuming work without pay also will help in preparing you for the reality of the immediate future.

Recognize this from the beginning: everything is going to take far longer and cost far more than you imagine, no matter how good your planning. Be especially attentive and realistic about acquiring adequate capital. One of the main reasons businesses fail is because they run out of money before they can achieve momentum and are unable to raise more.

If you are going to have partners, they should be people you have known for a long time, that you trust implicitly and care about. And you should make a deep commitment to one another at the beginning.

Lots of businesses break apart or fail because of differences between partners, oftentimes even when they are family members. Bill, Powell, and I pledged that we always would stand together, no matter what. No issues, major or petty, would be allowed to come between us. That was key for us, and it worked extraordinarily well.

You won't be able to choose investors as carefully as you can partners, but once you do, you must commit yourself to building the same compatibility and trust that you share with partners. That will take time, but it must be done. Differences of opinion will arise, but they must always be met with respect and civility so that bitterness, acrimony, and divisive feelings are never allowed to sprout and grow.

I said early in this book that business functions on relationships, and that's true between investors and managers, managers and employees, as well as with customers, suppliers, and others. It's always easier to accomplish things if you trust and respect the people with whom you are working.

Work hard from the beginning to build these necessary relationships. This will take a lot of time and effort, but it will be invaluable in the long run. Take advantage of crises in building relationships. If you can show somebody that you can help solve problems, work well under pressure, and be depended on in a pinch, it will seal trust.

Hiring is extremely important and should be done with great care. We decided at the beginning that this was a matter on which we would not compromise. We would search for the best people to fill the positions, even if we had to pay more to get them than we could afford at the time. We would give them the best tools to work with and the freedom for self-initiative. We also would provide incentives to reward them if we succeeded.

I think our decision about hiring was the most important one we made in building the company and ensuring its future. Most of the people we hired in our first years are still with us and playing major roles.

We also made certain that the management team that we built was congenial, united, and without internal political intrigue, which must be squelched before it has a chance to ferment and become destructive.

I know that some of the things I'm saying seem so obvious as to sound trite when the words are set down on paper, but they are far from that. Building an atmosphere for employees, a culture, if you will, that is open, honest, forthright, friendly, and accepting, and that sets standards for expectations and ethical behavior is absolutely essential.

We often get comments from visitors about what a warm and congenial place RF Micro is, and what a sense of family exists. We worked hard from the beginning to achieve that and strive to keep it going by making top managers accessible to all employees and by staging company events. We have cookouts where top managers do the grilling and serving. We have chili cookoffs, humorous roasts of top officials, and other fun events. We once bought out most of the seats at a Greensboro Bats baseball game and invited all of our employees and their families. After practicing for two days, Bill threw the first pitch—a strike, although some claimed the umpire was blind.

Let me briefly discuss another topic about which I frequently am asked by students. That is negotiating, an exercise in which I have spent untold hours in recent years.

What are the secrets of negotiating? I'm asked. Well, I wish I knew.

The truth is that there are none. Every situation is different, and has to be dealt with in particular and sometimes even peculiar ways. You sort of have to feel your way through it, and there's some art in that, I suspect, although I wouldn't be brash enough to claim any command of it.

I do have some thoughts about it, though. One is that if you are in a position of power, it's always tempting to take advantage, but if you do, it likely will come back to bite you somewhere down the road. And if you make a habit of it in life, you probably will die lonely and scorned.

In most negotiations, one party has something that another wants to one degree or another. The possessing party is apt to think that whatever it has is priceless, and the seeking party is going to hold it in somewhat lesser value. In a situation like this, no matter which side you're on, it's always easy to anger and offend, and that should be avoided for many reasons—not the least of which are your blood pressure and mental health.

The most important thing to remember is that in a truly successful negotiation, however you bring it about, each side should come away feeling that it got a good deal.

That brings me to a few final words, and once again they sound so obvious, and have been spoken so often—never more effectively, perhaps, than by Winston Churchill—as to seem to warrant not bothering to set them down. They are short and simple, and I offer them for the eternal verity that they hold.

Don't give up.

Be bullheaded in resolve. Keep butting. Try new things. Take risks. Be innovative. Force yourself to push ahead even when you're at your lowest and the future looks dark and hopeless. Just because the odds seem against you doesn't mean they can't be changed.

Not every new business will succeed, of course. Some people start several before they find their way.

My wish for all who hold the entrepreneur's dream is that it come as true for you as it did for Bill, and Powell, and me.

Index

4Kids Entertainment 178
A&T State University
  partnership with RF Micro 129

Adams, David 65
Adams, Kathy 88-90, 91
Adkins, Fred 87, 90, 176
Advanced Technology Ventures 24, 27, 33
Advertising campaign 69-74, 90
Agere Systems
  alliance with 183, 195-96

Alex Brown 107
Allen Telecom 29
Allen, George 193
Alliance Technology Ventures 107
Amazon.com 169
America On Line
  buys Time Warner 169

American Radio-Telephone 48
Analog Devices 9, 12, 14, 15-16, 20, 22, 49, 175
Andrews, Scott 55, 56
Applied Signal Technology, Inc. 21
Arthur Andersen 189
Asian stock market crash 148
AT&T 48, 93
  Dragon Phone 66-67, 74-79, 82, 87

Atmel 187
Bachert, Pete 80
Bardeen, John 45
Bell Laboratories 45, 48
Bell Telephone Company 42
Bell, Alexander Graham 41-42, 44
Benjamin, O.J. "Ben" 76
Bezos, Jeff 169
BiCMOS technology 198
Black, Fischer 193
Black-Scholes formula 193, 194
Bluetooth technology
  RF Micro expands into 199-200, 205-6, 211

Bowne and Company 134
Boxer, Barbara 193
Brantley Venture Partners 25, 33, 107
Brattain, Walter 45
Bridge City Tool Works 203-4
Briggs, Robert 23
Broadband technology 160
Bruggeworth, Bob 200, 212
Burr-Brown Corporation 30, 31, 32-33, 36, 58
Burson-Marsteller 75

Bush, Wes 161, 163, 164
Cain Technologies 22, 83
Cain-White & Company 22
Cambridge Silicon Radio 199, 205
Campbell, Doug 150
Cannino, Vinnie 72
Carolinas Capital 86
Catanoso, Justin 120, 201
Cecchini, Larry 71
Cecil, Mike 71, 90
Cell phones 12, 214
  business takes off 92
  early 47-48
  growing sales of 211
  growth of 151
  RF Micro power amps in majority of 210
  sales drop 178
  transmission standards 82-83

Cellular Technology Industry Association 48, 49
Chamis, Eleni 155
Channel Technology
  RF Micro purchase of 205

Chong, Kellie 58, 90, 175
Cincinnati Microwave 93
Clower, Jerry 104
CMAC 96
Coca-Cola 194
Code division multiple access (CDMA) 82
Collins Radio 155
Collins, Jim 21
Computer Labs 9, 10
ConCannon, Michael 159
Conexant
  RF Micro purchase of 198

Convergence 205
Cooper, Martin 48
Cordless phone technology 12
  expands with RF2103 chip 74-79

Corporate scandal
  impact on tech companies 189-94

Craver, Richard 175, 201
Creviston, Eric 176
Cypert, Ed 102
Dain Rauscher Wessels 150
Danielson, Merci 89
De Forest, Lee 43-44, 46
Digital Security Controls 51, 52, 64-65, 93
Digital technology 49

Dillingham, Jerry 72, 73
Dot.com bubble bursts 167-74, 190, 192, 194
Dreier, David 193
Duckworth, Linda 35, 88
Dunbridge, Barry 56
Dunn, Doug 91
Dzura, Dennis 21, 23
El Dorado Ventures 25
Electricity
    history of 40

Engel, Joel 48
Enron collapse 189, 191
Ensign, John 193
Enzi, Michael 193
Ericsson 88, 91, 160, 180
Eshoo, Anna 193
Evans, Matt 181, 186
Fairchild Semiconductor 45, 47
Faraday, Michael 40
Federal Communications Commission 46, 47, 48
Feller, Ben 156
Fessenden, Aubrey 46
Fiber-optics technology
    and gallium arsenide HBT chips 160
    collapse of market 165

Financial Accounting Standards Board 193-94
First Union 171
Fishman, Jerald 14
Flanigan, James 172
Fleming, Bob 86
Fleming, John 43
Flextronics 180
Frequency division multiple access (FDMA) 82
Garner, Andy 17
Garner, Lynn 11
Geiss, Art 122
General Electric 194
Gerding, Bruce 76
Gibson, Paul 170
Gigabit Logic 49, 54
Gillin, Eric 173
Glavin, Charlie 149, 150
Gleason, John 159
Global Positioning System (GPS)
    impact on RF Micro business 187

Global System for Mobile Communications (GSM)
82-83
Goldstar 110
Gray, Elisha 41, 42
Gray, Stephen 40
Grechkin, Igor 202
Green, Don 61, 62
Greenspan, Alan 212
Griffin, Curtis 80-82, 88, 94, 95
Gruber, Werner 102
Hambrecht & Quist 132
Hanneman, Tim 98, 102, 162
Hansen, Ken 80
Harwood, Beth 74, 79
Harwood, John 74
Henry, Joseph 40
Hertz, Heinrich 42
Hicks, Bob 58, 60, 87, 90, 175
Howard, Jill 58
Howland, Jeff 164, 166
Hunt, James 120, 121, 123
Hypnarowski, Joe 85
Hyundai 143
IBM 153, 159, 160, 183, 187, 195
    considering buyout of RF Micro 132-33

Ibuka, Masaru 46
Infenion 159
Integrated circuitry 12, 47
Intel Corporation 45, 66

Iraq War
    impact on business 206

Itron 92
ITT 175
Jackson, Thomas Penfield 172
Jazz Semiconductor
    RF Micro purchase of 198

Joyner, Amy 201
Jubak, Jim 177
Kelly, Kevin 22, 83, 127
Kelly, Mervin 45
Kemp, George 43
Kilby, Jack 47
Kitty Hawk Capital 25, 26, 28, 37, 107
Krabbe, Hank 58
Krouse, Peter 136, 140, 154, 156, 158
Lehman Brothers 208
Leiwe, Dan 96
Levin, Carl 193
Lofman, Johan 112, 115
Lucent Technologies 183
Lyon, Dave 206
Malkemes, Bob 74, 77
Marconi, Guglielmo 42-43
Marconi's Wireless Telegraph Company 42
Maxwell, James Clerk 42
May, Samuel 171
McCain, John 193
McKechnie, Mark 181
Menlow, David 136
Merrill Lynch 212
Merton, Robert 193
Microsoft
    violation of anti-trust laws 172

Miller, Karl 161
Misenheimer, Milton 12
Moeller, Jim 181
Montgomery Securities 132
Moore, Elizabeth 175
Moore, Forrest 175
Morita, Akio 46
Morse, Samuel 40-41
Motorola Corporation 48, 93, 143, 180, 211, 213
    emergency two-way radios 80-82, 88
    problems with two-way radios 94-96

Nadler, A.J. 176, 199
Nationsbank 108
Neal, Albert 13, 16, 39, 50, 124
Neal, Annette 202
Neal, Jerry D. 29, 57, 86, 89, 90, 115, 143-44, 154, 204, 214
    gets laid off from Analog Devices 16-18
    love of radio 13
    marketing experience 15

Neal, Linda Stewart 65, 89
Nicol, Alan 65
Nippondenso 51, 60, 64, 143, 213, 214
Nokia 88, 91, 93, 101, 110, 126, 136, 143, 145, 147, 153, 154, 177, 180, 209, 212
    pressures RF Micro to establish own fab 112
    secrecy agreement 135

Norbury, David 64, 69, 110, 162
    retirement of 196, 200-204

Norman, Randy 96
North Carolina Enterprise Fund 86
Northrup Grumman 188
Norwest Venture Capital 86
Noyce, Robert 47, 66
Oersted, Christian 40
Oki, Aaron 56, 91
Oliver, Clarence 65

Ooi, Joey 80
Oppenheimer & Company 132
Optimum technology matching 69-70, 84
PaineWebber 171
Paladino, Albert E. 24-25, 26, 27, 28, 29, 30, 33,
    34, 57, 111, 112, 113, 114, 115, 119-20, 160,
    162, 163, 197, 200, 204
Paul, Robert 29, 30
Pawlik, Mike 30, 31, 35, 36
PCS phones 84, 96, 101
Perkins, Anthony 173
Peterson, John 51
Petroni, Lou 114
Pfau, Dale 146
Pitt, Harvey 190
Polaris technology 213, 214
Powell, Bonnie 13
Powers Group 71
Pratt, Bill 9-14, 29, 53-55, 57, 59, 61, 62, 86, 90,
    144, 154, 162, 197, 214
    on the role of engineers 50

Pratt, Jeanne Anne 25
Pratt, William H. 58, 91
Priddy, William A. "Dean" 35
PrimeCo 101
Qualcomm 83-85, 136, 169, 177
    alliance dissolves with 180
    and specifications issues 96-98, 100-106, 109
    dropping as a customer 126-28
    investing in RF Micro 107
    returns as a customer 170-71

Quality monitoring program for foundries 85, 87
Radio
    invention of 42-43

Recession 180, 184, 197
Reichard, Peter 121
Reid, Harry 193
Resonext
    Moscow software division 201-2
    RF Micro purchase of 200, 211

Reynolds, Leonard 65, 90
RF CMOS technology 198
RF Micro Devices
    and naming chips 66
    builds automated packaging plant 158-59
    builds Fab Three foundry 179
    builds new foundries 151, 157-58
    builds new headquarters 117-18, 129-30,
        151, 174
    builds first semiconductor foundry 111-20,
        122-24, 140-44
    employee turnover rates 192
    expands foundry 147
    expansion plans 68, 99-100
    explores cell phone market 48-49
    finding funding for 20-38
    fire at foundry 149
    first press conference 75-77
    formation of 14, 49
    hiring philosophy 175-76
    investment in 33, 37, 68, 86, 107-8
    launches new R&D department 182
    named number two fastest growing company
        178
    named one of 500 fastest growing companies
        148
    named second fastest growing tech company
        in North Carolina 106
    new rounds of financing 86
    opens broadband division 164-65
    opens design centers 155
    opens largest molecular beam epitaxy fab 156
    opens new design center in Denmark 178
    opens new sales center in Japan 197

opens new sales center in South Korea 196
opens new testing lab 165
opens sales center in Taiwan 178
plans new plant in China 178
public stock offering 107, 121, 122, 131-40
restructuring of 211-12

RF Nitro Communications
    acquired by RF Micro 186

RF2103 chip
    and AT&T's Dragon Phone 66-67, 74-79, 82, 87
    and Motorola's two-way radios 80-82, 88

RF2115 chip 88, 94
Rider, Brent 25
Rockwell International 54-55, 110, 155, 198
Rodgers, T.J. 191
Rudy, Suzanne 164, 166
Rund, Raymond J. 25, 33, 37, 57, 64
Samsung 110, 196, 211
Sarbanes-Oxley Act 190, 193
SARS epidemic
    impact on business 206, 208

Scalf, Jeffrey 65
Scholes, Myron 193
Schrock, Don 109, 170, 198
Scism, Jack 70
Semiconductor chips
    gallium arsenide 53
    gallium nitride 197, 214
    HBT 53, 54, 84
    MESFET 53, 54, 84

Semiconductor technology
    invention of 45

September 11 attack
    impact on business 184-86

Seymour, Powell 12-14, 29, 60, 61, 62, 86, 87,
    90, 115, 154, 214
Seymour, Wendy 58
Shah, Manish 136
Shealey, Jeff 186
Shealey, John 186
Shelton, Ralph 121
Shockley, William 45, 53
Shrock, Don 127
Silicon germanium technology 159
Silicon technology
    increasing part of RF Micro's business 182
    RF Micro explores 153

Silicon Wave
    RF Micro purchase of 205-6, 211

Soicher, Ron 159
Sony Electronics 46, 84, 97
Sprint 101
ST Microelectronics 159
Stata, Ray 16, 22, 49
Steel, Vic 79, 92, 176, 182
Stewart, Gene 65
Stewart, Martha 190
Stock options
    impact of expensing on tech companies 191-94

Stone Communications 138
Sugar, Ron 161
Suitt Construction Company 141
Sulpizio, Rich 109, 110
SVM Star Ventures 107
Technology industry
    and 1987 recession 13

Telegraph

    invention of 40-41

Telephone
    creation of transcontinental calls 44
    invention of 41-42
    mobile 46-47, 51

Texas Instruments 47, 54
Thompson, Greg 92, 176
Time division multiple access (TDMA) 82
TRW 55, 56, 147, 165, 166, 169, 171, 172, 187-88
    and broadband license 160-61, 163-64
    and production issues 94, 98-99, 106, 108, 110
    licenses HBT technology to RF Micro 111-16
    taking in other micro-electronics companies 133
    unloads RF Micro stock to buy LucasVarity 155-56

TSMC 213
Tuen, Steven 136, 137
Van Buskirk, Bob 91, 98, 102, 110, 111, 133, 161
van der Kaay, Erik 29, 30, 34, 57, 108, g160,
162, 206
Vandervoet, Dave 98, 102, 117
Vargo, Ron 161
Venture capital 107, 168, 191
    and RF Micro 23-32, 37-38

Volta, Allesandro 40
Watson, Thomas 42, 44
Western Electric 42
Western Union Company 41
White, Harvey 100, 101
Wilkinson, Walter 25, 26, 28, 29, 33, 34, 37, 57,
109, 111, 112, 113, 114, 115, 162
WLAN technology
    RF Micro expands into 199-200

Yau, Michael 51
Zhu, Theodore 198
Zimmer, Patric 157
Zinkiewicz, Terri 162

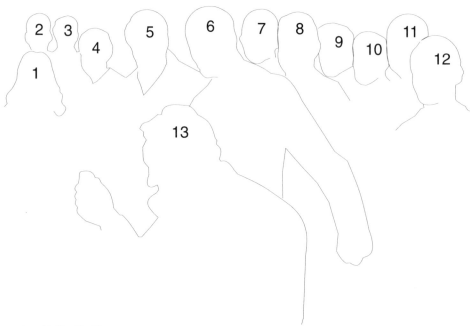

1. Kellie H. Chong
   Director, Mobile Applications
   Product Line

2. William A. Priddy, Jr.
   Chief Financial Officer,
   Corporate Vice President of
   Administration and Secretary

3. Bob Hicks
   General Manager,
   Global Operations

4. Leonard D. Reynolds, Jr.
   Principal Engineer

5. Powell T. Seymour
   Co-Founder,
   Corporate Vice President of
   Strategic Operations and
   Assistant Secretary

6. Jerry D. Neal
   Co-Founder,
   Executive Vice President of
   Marketing and Strategic
   Development

7. Dr. Albert E. Paladino
   Original Investor,
   RFMD Chairman of the Board

8. William J. Pratt
   Co-Founder,
   Chief Technical Officer,
   Corporate Vice President

9. Erik van der Kaay
   Original Investor,
   RFMD Board of Directors,
   Retired Chairman of the Board,
   Symmetricon, Inc.

10. Walter H. Wilkinson, Jr.
    Original Investor,
    RFMD Board of Directors,
    Founder and General Partner,
    Kitty Hawk Capital

11. David A. Norbury
    RFMD Board of Directors,
    Retired Chief Executive Officer, RFMD

12. William H. Pratt
    Senior Engineering Manager

13. Alexander Graham Bell
    The Father of Modern
    Telephone Service